The Anthropocene and the Humanities

# THE ANTHROPOCENE AND THE HUMANITIES

FROM CLIMATE CHANGE TO A NEW AGE OF SUSTAINABILITY

● ● ●

CAROLYN MERCHANT

Yale
UNIVERSITY PRESS

New Haven and London

Published with assistance from the foundation established in memory of Calvin Chapin of the Class of 1788, Yale College.

Copyright © 2020 by Carolyn Merchant. All rights reserved. This book may not be reproduced, in whole or in part, including illustrations, in any form (beyond that copying permitted by Sections 107 and 108 of the U.S. Copyright Law and except by reviewers for the public press), without written permission from the publishers.

Yale University Press books may be purchased in quantity for educational, business, or promotional use. For information, please e-mail sales.press@yale.edu (U.S. office) or sales@yaleup.co.uk (U.K. office).

Set in Adobe Garamond and Gotham types by Tseng Information Systems, Inc.
Printed and bound by CPI Group (UK) Ltd, Croydon, CR0 4YY

Library of Congress Control Number: 2019948080
ISBN 978-0-300-24423-6 (hardcover : alk. paper)

A catalogue record for this book is available from the British Library.
This paper meets the requirements of ANSI/NISO Z39.48-1992 (Permanence of Paper).

10 9 8 7 6 5 4 3 2 1

For my family

# Contents

| | |
|---|---|
| Preface | ix |
| Acknowledgments | xv |
| Introduction: Climate Change and the Anthropocene | 1 |
| 1 History | 26 |
| 2 Art | 46 |
| 3 Literature | 66 |
| 4 Religion | 90 |
| 5 Philosophy | 107 |
| 6 Ethics and Justice | 127 |
| Epilogue: The Future of Humanity and the Earth | 144 |
| Notes | 157 |
| Bibliography | 173 |
| Illustration Credits | 195 |
| Index | 201 |

# Preface

We are all blips of life in a sea of eternity. Each life, whether human, animal, plant, or bacterium, constantly combats the most relentless and unforgiving of nature's laws—the second law of thermodynamics. Each life loses its battle. Why? Because the world is running down. It is moving from order to disorder, while entropy—the energy unavailable to perform useful work—is constantly increasing. The end of the earth will be a heat death in which all temperatures are equal. No movement, no change, no transformation.

Evolution, which creates new order, at first seems to defy entropy. It creates ever more complex varieties, species, genera, and families as new energy is supplied from the sun. But as each new life-form confronts the second law, it loses its battle. Each life reproduces its own kind, grows old, and dies. Each new life is but a blip that exists for a brief moment in time before it vanishes forever. Whether the universe itself will succumb to a heat death as it constantly expands and its temperature differential approaches zero or whether it will contract and collapse into a black hole and reemerge is an open question.

But as we approach the mid-twenty-first century, losing hundreds of life-forms is of mounting concern. As we face the effects of global warming from the continued burning of fossil fuels, we will increasingly encounter extreme weather patterns, melting polar ice, hurricanes, floods, and tornadoes, raising the specter of the death of the earth itself. The concept of a new geological epoch, the Anthropocene (as named by scientists Paul Crutzen and Eugene Stoermer in the year 2000), in which human activities impact nature in devastating new ways, suggests that the earth—*as we know it today*—may cease to exist in the future.[1] Rethinking nature in the Anthropocene—the period from the advent of the steam engine in the late 1700s to today's increased burning of fossil fuels (coal, oil, and natural gas)—has profound implications for reconceptualizing, not only the sciences, but the humanities themselves. How, for example, is the air and water pollution associated with global warming reflected in history, art, literature, religion, philosophy, ethics, and justice? How is the very idea of what it means to be human altered with climate change? What does the future hold for humans and the humanities in the Age of the Anthropocene?

Swedish environmental historian Sverker Sörlin, in "Environmental Turn in the Human Sciences," writes: "Considerable energies are going into the emerging concept of *environmental humanities*. This is a broad multidisciplinary approach that signals a new willingness in the humanities ... for a common effort in which the relevance of human action is on par with the environmental aspect. Programs or other initiatives for the environmental humanities have already started to emerge in universities in Europe, Australia, and the United States, including Princeton, Stanford, and UCLA."[2] Although numerous articles and books have been written about the Anthropocene in the fields of the sciences, politics, economics, and governance, to date there are relatively few analyses of the Anthropocene as it relates to the humanities.

In the Age of the Anthropocene, much is at stake for the mutual survival of humanity and nature. The humanities are of critical importance in bringing the environmental crisis of the twenty-first century to the attention of the American public. Engaging individuals, governing bodies, and communities in implementing change requires not only the research of scientists but equally, perhaps especially, the insights of humanists. A theoretical framework for the environmental humanities that is applicable to large-scale, complex environmental problems in the Anthropocene must be sought.

The idea of the Anthropocene can help us reconceptualize the humanities in new ways that make them compelling for the twenty-first century. Language and images can play formative roles in creating awareness and changing personal behavior and public policy. The humanities (as exemplified by history, art, literature, religion, philosophy, ethics, and justice) can create new and compelling cognizance of the critical choices facing us during the next fifty to one hundred years and beyond.

This book is meant for an educated public interested in the current state of the planet, its future, and what we as humans can do to preserve life on earth. It can be used in undergraduate courses and graduate seminars that focus on the environment, the humanities, and social sciences as well as in e-book clubs and discussion groups. It is meant to provoke thoughtful responses and inspire creative solutions by examining the arts and humanities, science and history, ethics and justice.

In the following chapters, I introduce the term *Anthropocene*, ask why this new designation matters, and critically assess the various meanings and significance attributed to it by scientists and humanists. I argue that the concept of the Anthropocene goes beyond earlier concepts and periodizations such as preindustrial, colonial, industrial, modern, and postmodern by presenting a clear and forceful characterization of the future crisis humankind faces. I examine ideas associated with the origins of the era of the Anthropocene, especially in

Western culture, and what precepts can point toward a new era of sustainability based on energy, process, and "green" science. I show how and why the connection of the Anthropocene and the humanities matters for all of us in the future.

I focus the chapters of this book on Europe, particularly England, and the United States, where industrialization began, but I also suggest areas and continents where the idea of the Anthropocene can and should be further explored. My goal throughout is to examine problems with Western histories and ideas and to propose new principles of process and partnership that can become ideals for the future. In so doing, I draw on and synthesize ideas that span the length of my academic career. I include concepts from my books and articles that offer insights into what the Age of the Anthropocene is, how it is exemplified, and the possibility of transforming it into an Age of Sustainability.

This book does not claim to be comprehensive. I make no attempt to cover all countries, continents, and periods or to cite all the books on the Anthropocene that have appeared in recent years. Rather, I choose thought-provoking examples to gain insights into the relationship between the Anthropocene and the humanities and indicate where research can fruitfully be expanded by others. To make the book more accessible to a general audience, I include images of the women and men who played major roles in the emergence of the Anthropocene as well as the art that characterizes its significance and consequences.

It is crucial for the future of humanity that we explore the causes and consequences of, and intimate connections with, climate change in the Anthropocene. The increasing buildup of greenhouse gases, global warming, and the melting of Arctic, Antarctic, and mountaintop glaciers and snows has an enormous impact on rising sea levels and hence on life-forms around the world. The effects of climate change are manifested in warming waters, drought, desertification, extinction, and the migration of species. Human populations are likewise adversely affected. Women, especially those in developing countries, experience the burden of increased work, such as carrying water from distant sources, gathering fuel, and caring for families. The additional labor results in suffering and loss of life, especially among the poor, working classes, racially discriminated peoples, and those of the female gender.

It is critical to reach solutions to our global ecological and humanitarian crisis. Inspired by the humanities, such solutions will entail new scientific approaches, technologies, politics, ethics, and especially changes in the class, race, and gender differentials that result in suffering for so many people. The earth itself will continue in some form, but perhaps in a much altered state. It is incumbent on those of us living now to make the changes that will save humanity and nature as we know them today.

We are all visitors on the earth.

# Acknowledgments

I wish to thank the many people who have contributed ideas and conversations to the materials that make up *The Anthropocene and the Humanities*. I am particularly grateful to Jennifer Wells, my coauthor on an earlier work, "Melting Ice: Climate Change and the Humanities," that appeared in *Confluence* (2009), portions of which are included here. I likewise appreciate comments offered by University of California, Berkeley (UCB), colleagues Carolyn Finney, Robert Hass, Alastair Iles, Rachel Morello-Frosch, Garrison Sposito, Kimberly TallBear, and David Winickoff. UCB students Marley Pirochta and Rachel Rombardo provided invaluable assistance with manuscript preparation and permissions through a grant from UC Berkeley, College of Natural Resources, Sponsored Projects for Undergraduate Research (SPUR) in the fall of 2018.

An award from UC Berkeley's Townsend Center for the Humanities provided support for a course in the spring semester of 2016: "The Fate of Nature in the Anthropocene." Six faculty members and twelve graduate students read numerous books and articles and met weekly for robust discussions about the concept of the Anthropocene and its impact on the environment and humanity. Research support was also provided by a Futures Grant from the University of California at Berkeley and a fellowship at the Center for Advanced Study in the Behavioral Sciences (CASBS), Stanford University, in the fall semester of 2017. I thank the members of the class of 2017 for insightful and inspiring conversations and everyone at CASBS for assistance in obtaining books and articles and providing a congenial place to read, think, and write.

This book draws on and synthesizes ideas from my earlier books, especially *The Death of Nature* (1980; 2e, 1990; 3e, 2020). In it I discussed the transition from the living, organic world of the sixteenth century Renaissance, in which the earth was a nurturing mother, to the mechanistic world of the seventeenth century, in which matter was dead and inert and God was an engineer, mathematician, and clockmaker. In the present book, I refer to a "second death of nature" in the Anthropocene—the period from the invention of James Watt's steam engine in 1784 to the present—which allowed greenhouse gases to accumulate in the atmosphere resulting in "climate change." I also integrate ideas from my other books, bringing together concepts from history and new ideas that help us look toward the future and toward a New Age of Sustainability.

My colleagues and former students Kenneth Worthy, Elizabeth Allison, and Whitney A. Bauman have published a volume on my work titled *After the Death of Nature: Carolyn Merchant and the Future of Human-Earth Relations* (Routledge, 2019). I am honored and extremely appreciative. I thank them and Routledge for granting permission to use portions of my "Afterword" in this book. I am likewise grateful for the helpful suggestions offered by the reviewers of this book, Edward Melillo, Mary Evelyn Tucker, and an anonymous reviewer. I especially thank my editor at Yale University Press, Jean Thomson Black, her assistant, Michael Deneen, production editor Jeffrey Schier, and indexer Fred Kameny for their excellent help in preparing the book for publication.

Most of all, I thank my husband, Charles Sellers, for stimulating conversations, ideas, and moral support during the research and writing of this book.

The Anthropocene and the Humanities

INTRODUCTION
# Climate Change and the Anthropocene

Climate change is the most critical issue for the long-term well-being of humanity in the twenty-first century. Scientists now broadly agree that anthropogenic or human-driven inputs exacerbate climate change and that a wide range of strategies to manage its effects are possible. But bringing the implications of global warming and potential resolutions to the American public requires the collaboration not only of scientists but also of humanists. Moving beyond the sciences, we should take note of human contributions to climate change and study the ways in which the humanities can and should engage with this complex field.

## The Anthropocene

In their foundational one-page paper "The Anthropocene," published in 2000, scientists Paul Crutzen and Eugene Stoermer introduced the concept of the Anthropocene—that is, the age of humanity—and directly linked it to the human (anthropogenic) causes of climate change. Crutzen, a Dutch atmospheric chemist at the Max Planck Institute for Chemistry in Germany, won a 1995 Nobel Prize in chemistry for research on the ozone layer. Stoermer, professor of biology at the University of Michigan, first coined and used the term *Anthropocene* in the early 1980s to refer to human impacts on the planet. But it was with their joint paper in 2000—the

INTRODUCTION

*Figures I.1 and I.2. Paul Crutzen (b. 1933) and Eugene Stoermer (1934–2012)*

first year of the new millennium—that the term caught on. Since then, a host of books and articles on the implications of the Anthropocene for many fields have appeared.[1]

When did the Anthropocene actually begin? Crutzen and Stoermer wrote, "To assign a more specific date to the onset of the 'anthropocene' seems somewhat arbitrary, but we propose the latter part of the 18th century.... Such a starting date also coincides with James Watt's invention of the steam engine in 1784."[2] The emphasis by Crutzen and Stoermer on the 1780s is critical because that was the era when the burning of fossil fuels in steam engines made possible subsequent inventions, such as steamboats, trains, and steam-driven industries, thereby escalating the discharge of greenhouse gases into the atmosphere. The Anthropocene, according to Crutzen and Stoermer, was marked by the substantial increase of greenhouse gases from the burning of fossil fuels in the late eighteenth century, when "data retrieved from glacial ice cores show the beginning of a growth in the

## INTRODUCTION

atmospheric concentrations of several 'greenhouse gases,' in particular $CO_2$ [carbon dioxide] and $CH_4$ [methane]."[3]

What are the implications of the Anthropocene for the future? As Crutzen and Stoermer noted: "In a few generations mankind is exhausting the fossil fuels that were generated over several hundred million years." The most important step in preserving the earth for the future will be for scientists and engineers to work together with the public to develop a strategy of "global sustainable, resource management."[4]

The Anthropocene, according to Crutzen and Stoermer, follows the Holocene, the post-glacial epoch of 10,000–12,000 BP (before the present), when human activities first became a significant force on the earth. At that time, there were approximately 5 million people in the world. Known as the interglacial warm period, the Holocene had a relatively stable climate, allowing for the establishment of human settlements around the globe and therefore the development of crop agriculture, including wheat, oats, barley, rice, sorghum, corn, beans, and squash. Coupled with horticulture was the domestication of animals such as cows, pigs, sheep, goats, and horses (Fig. I.3).

Then, between 1 CE and the present, human population growth

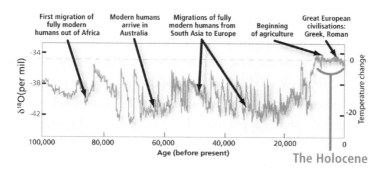

*Figure I.3. The Holocene*

began to accelerate, increasing from 200 million in 1 CE to 500 million in 1650, 1 billion by 1850, 2 billion by 1930, 6 billion in 1999, and a projected 8 billion by 2024. Global temperatures rose steadily between 1880 and 2010. Average global temperatures were cooler than the mean of 0° in 1940 and warmer from the 1940s to the present (Fig. I.4). The average surface temperature of the earth during the twentieth century was 13.7°C (Celsius), or 56.7°F (Fahrenheit). To keep global temperature increases below 2°C (3.6°F)—the temperature that will support life in the coming decades—will require major decreases in fossil fuel emissions and increases in forests, grasslands, wetlands, and farmlands.[5]

The "human footprint" on the planet can be illustrated by several graphs, all of which show exponential growth (Fig. I.5). Carbon dioxide ($CO_2$) concentrations between 1750 and 2000 rose from around 50 parts per million (ppm) by volume to 360 ppm. By the year 2000, the damming of rivers rose from virtually no dams in the late nineteenth century to twenty-five thousand. While human-driven extinctions of living organisms have been noted for hundreds of years, the rate of extinctions around the globe rose by the late nineteenth century to around thirty thousand (and is projected to become earth's sixth great extinction event), while land use by humans increased from around 10 percent of the earth's surface in 1900 to over 25 percent.[6]

What do these trends mean for the future of humanity and the planet? Data from the Environmental Protection Agency (EPA) for January 2017 between the years 2000 and 2100 show *projected* atmospheric greenhouse gas concentrations. The "highest emissions pathway" for 2100 shows approximately 1,300 ppm of $CO_2$ equivalent. The "higher pathway" shows approximately 800 ppm, while the "lower pathway" predicts around 600 ppm and the "lowest" peaks around 2040 at circa 450 ppm and then declines to just above 400

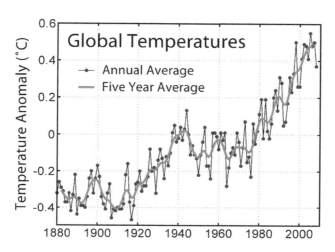

Figure I.4. Global temperature change, 1880–2010

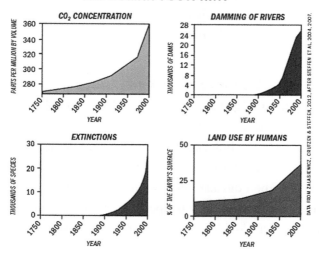

Figure I.5. The human footprint

## INTRODUCTION

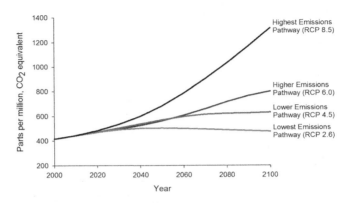

*Figure I.6. Environmental Protection Agency's projected atmospheric greenhouse gas concentrations, 2000–2100*

ppm by 2100 (Fig. I.6). The organization 350.org was formed in 2007 to pressure the nation to create policies to reduce the parts per million of $CO_2$ in the atmosphere from 400 ppm to 350 ppm as the safe upper limit for life on the planet.[7]

### History of Climate Change

The concept of global warming (now known as climate change) was first proposed by Swedish scientist Svante Arrhenius in an 1896 paper titled "On the Influence of Carbonic Acid in the Air upon the Temperature of the Ground." Arrhenius argued that the continued addition of carbon dioxide into the earth's atmosphere might result in an increase in the earth's temperature. If the amount of atmospheric $CO_2$ were doubled, the earth's surface temperature could rise by 5 percent.[8]

Warnings of what later came to be known as the "greenhouse effect," however, fell on deaf ears. The subject received little publicity

INTRODUCTION

*Figure I.7. Svante Arrhenius (1859–1927)*

and did not garner immediate approval by scientists. In the 1940s and 1950s new research seemed to indicate that the oceans could absorb $CO_2$, thus mitigating the impacts of climate change, and that there might even be a cooling trend on the planet. International Geosphere-Biosphere Programme (IGBP) data show that a "great acceleration" in human-induced changes to the earth's systems occurred during the 1950s.

It was not until the 1980s that scientific consensus began to build that the climate had warmed significantly since the 1860s. Global warming was termed the "greenhouse effect" because of the rise of carbon dioxide ($CO_2$), methane ($CH_4$), nitrous oxide ($N_2O$), hydrofluorocarbons (HFCs), perfluorocarbons (PFCs), sulfur hexafluorides ($SF_6$), and other "greenhouse gases."[9]

The IGBP was initiated in 1987 for the purpose of gathering and coordinating research on biological, chemical, and physical processes and their interactions with human social and economic processes on the earth's surface. The goal was to help provide a way forward to

INTRODUCTION

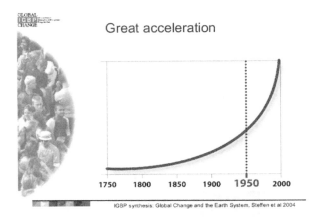

*Figure I.8. IGBP "great acceleration" data*

reduce greenhouse gases so that the planet could achieve sustainability.[10]

In 1988, the Intergovernmental Panel on Climate Change (IPCC), consisting of some 2,500 scientists from 60 countries, was founded by the United Nations Environmental Programme. The scientists released reports of their findings beginning in 1992 with continuing updates until 2014. The Kyoto Protocol, which was formulated in 1998 and signed in 2001, committed some 186 countries to reduce carbon emissions from 1990 levels by 5 percent by 2012 (see below). In the 2014 Fifth IPCC Assessment Report, the IPCC stated that human influence has been the most likely cause of observed warming since 1950 and that the level of confidence has increased since the Fourth Report. Moreover, the longer it takes to reduce our emissions, the more expensive it will be. The Sixth Report is due in 2022.[11]

But a report released on October 8, 2018, by the IPCC was even more ominous, moving the date of extreme danger to 2030: "Earth is on track to face devastating consequences of climate change—extreme drought, food shortages and deadly flooding—unless there's an 'unprecedented' effort made to reduce greenhouse gas emissions by 2030." It warned that "we must take 'rapid, far-reaching and unprecedented changes in all aspects of society' in order to save our planet."[12] In light of such dire predictions, how can the humanities help to spread information and clarify the consequences for humanity in the future?

### Climate Change and the Humanities

Scientific issues and debates about climate change have set up possibilities for responses by the humanities. Humanists—ethicists, writers, poets, artists, and theologians—are engaging with issues surrounding global climate change and its effects on peoples of different race, class, and gender. They are asking questions such as: what is nature, what does it mean to be human in the age of warming, and how can humans use technology to adapt to the new world? Both scientists and humanists can engage in climate change debates to illuminate options for the future.

In his 1989 book, *The End of Nature,* environmental writer Bill McKibben argued that no area of the earth today remains untouched by human pollution, including the atmosphere of the Arctic. First Nature (that is, evolved, prehuman nature) has been totally subsumed by humans and the human artifacts of Second Nature (or commodified nature).[13] Global warming calls for new and plural understandings of nature, the nature-culture web, and the techno-nature that follows.

Five overlapping themes offer insights into the humanistic dimensions of climate change: climate change and the arts, climate change and literature, climate change and religion, climate change and philosophy, and climate change and ethics/justice. Scientists, historians, artists, writers, philosophers, and theologians have made significant contributions. A theoretical framework for the environmental humanities and new theories of ethics and justice are applicable to large-scale, complex environmental problems.

The Earth Charter of 2000 states: "We stand at a critical moment in Earth's history, a time when humanity must choose its future.... In the midst of a magnificent diversity of cultures and life forms we are one human family and one Earth community with a common destiny.... It is imperative that we, the peoples of the Earth, declare our responsibility to one another, to the greater community of life, and to future generations."[14]

Overall, the cross-cutting theme of the humanistic dimensions of climate change can help individuals to resolve personal dilemmas and formulate individual ethical responses, while governing bodies can begin to formulate political responses to the implications of climate change for the sake of humanity and the future of the planet.

### Climate Change and Politics

What efforts have been initiated both globally and locally to reduce the levels of greenhouse gases? As a result of the Conference on Global Warming held in Kyoto, Japan, in 1997, the Kyoto Protocol was drafted with the goal of cutting emissions of greenhouse gases by 5 percent of 1990 levels by 2012. Revised in Brussels in 2001, the protocol was ratified in 2005 by thirty of the world's industrialized nations, with the United States and Australia declining to join unless developing nations were included in the targets. Within the U.S., however, California has assumed a leadership role with the passage of the Global Warming Solutions Act of 2006 (AB32), which seeks to reduce greenhouse gas emissions to 1990 levels by 2020. In September 2007, California's then governor, Arnold Schwarzenegger, addressed the United Nations on the urgency of responding to global climate change. Since then, Governor Jerry Brown has assumed the mantle of dealing with the global dimensions of climate change and brought worldwide attention to California's commitment to arrest global warming.[15]

Scientific consensus on the severity of unmanaged climate change is well established, but the ways and extent to which global warming can be mitigated are still under discussion. Recent manifestations of global warming include the bleaching of one-half of the world's coral reefs (1998); devastating droughts and floods throughout much of the world (1995–2018); and the fact that (to 2019) the year 2016 was the hottest in recorded history. In addition, some events have occurred that seem worse than most climate scientists had predicted, owing to multiple "feedback effects" leading to a marked increase in the speed of glacial melting. In July 2017, an iceberg the size of Luxembourg broke off the Antarctic ice shelf. It now appears that the fabled northwest shipping passage through the Arctic will soon become a reality. Circumpolar countries are vying for rights to oil reserves under the melting ice. And in late 2018, researchers announced that fossil fuel emissions had begun to rise again after a three-year lull.[16]

To save the earth as we know it today, it is critical to reduce carbon dioxide emissions through conversion to renewable energy sources and then to remove as much carbon as possible by planting forests and crops and through technologies such as underground carbon capture. If unmitigated, atmospheric carbon could affect the planet for thousands of years. While about one-third of the carbon dioxide emitted daily by fossil fuels is absorbed by oceans, this process gradually increases ocean acidity and inhibits the growth of corals and shell-forming organisms. More carbon is removed through much slower processes such as rock formation.

Numerous methods and technologies for humans to remove $CO_2$

have been proposed. Conserving and planting forests to covert $CO_2$ to oxygen ($O_2$) is essential. Beyond this, carbon engineering technologies include large structures built in plains and deserts that capture carbon directly from the atmosphere; the use of chemicals that absorb carbon; carbon purification and liquefaction; carbon capture and burial techniques; and the development of catalysts that accelerate natural weathering. The biggest question involves the speed and scale on which changes can be made. Without immediate and drastic changes that will move us to an Age of Sustainability, the Anthropocene could continue for hundreds of years.[17]

Responding to global warming is now considered by many to be "the moral imperative of our time." With the release of Al Gore's film *An Inconvenient Truth* in the summer of 2006, public concern skyrocketed, and in 2007, both Gore and the IPCC were awarded the Nobel Peace Prize. Gore characterized the problem as "a moral and spiritual challenge to all of humanity." In 2016, nearly half of U.S. adults recognized that climate change is due to human activity, and majorities in forty countries considered global warming a very serious issue. Environmental programs and courses on college campuses are overflowing with students who wish to be informed on issues and approaches to the resolution of climate change for their own futures and those of their children and grandchildren.[18]

Scenarios for dealing with global warming range from continuing business as usual to taking drastic measures now, before the effects become irreversible. Doubters such as Bjørn Lomborg argue that imposing limits on greenhouse gas (GHG) emissions is ineffective and too costly, while Ted Nordhaus and Michael Shellenberger maintain that it is politically unfeasible to try to persuade Americans to adopt changes.[19] Nevertheless, of people surveyed by Yale climatologist Anthony Leiserowitz, 67 percent strongly favor "requiring automakers to increase the fuel efficiency of cars, trucks, and SUVs to 35 miles per gallon, even if it means that a new car would cost up to $500 more to buy"; 64 percent favor "requiring any newly constructed home, residential, or commercial building to meet higher energy efficiency standards"; 55 percent strongly favor "requiring electric utilities to produce at least 20% of their electricity from wind, solar, or other renewable energy sources, even if it cost the average household an extra $100 a year"; and 42 percent strongly favor an international treaty that requires the United States to cut its emissions of carbon dioxide by 90 percent by 2050.[20]

## Alternatives to the Anthropocene

Since Crutzen and Stoermer's article was published in 2000, questions about the meaning of the term *Anthropocene* have been raised and intensely debated. These include: what exactly is the Anthropocene, what should it be called, when did it actually begin, and what does it mean for the future? The word *anthropos* (ἄνθρωπος) is the Greek word meaning man or human. The equivalent Latin word is *homo*. *Homo sapiens* is the genus of hominids that means modern humans who appeared in Africa around two hundred thousand years ago. The Anthropocene means the age of man or, more appropriately, the age of humanity.

Historians have placed the onset of the Anthropocene at different times in human history, such as the destruction of large mammals at the end of the last ice age, the onset of agriculture, the preindustrial era of the eighteenth century (proposed by Crutzen and Stoermer), the rise of industrial capitalism in the nineteenth century (the Capitalocene), the post–World War II nuclear age, or the "great acceleration" of the 1950s. If, for example, the Anthropocene basically refers to human impacts on the earth, then the human hunting of large mammals coupled with the burning of landscapes around fifty thousand years ago can be considered as a starting point. Another starting point is proposed by paleoclimatologist William Ruddiman, who states that the Anthropocene began with the clearing of forests for settled agriculture around 8000 BP. In this scenario, the spread of agriculture from the domestication of crops such as corn, beans, squash, wheat, rye, and rice, the domestication and disbursement of animals (cows, pigs, horses, goats, and sheep), and the impact of European germs on indigenous Americans are all vividly portrayed by Alfred Crosby in *The Columbian Exchange* (1973) and by Jared Diamond in *Guns, Germs, and Steel* (1997). From my perspective, however, the Anthropocene (as originally proposed by Crutzen and Stoermer) begins in the late eighteenth century when the emission of fossil fuels begins to change the atmosphere, leading to climate change.[21]

Still another proposed starting point for the Anthropocene, however, is the period of the 1950s. In 2002, Paul Crutzen and coauthors Will Steffen and John McNeill argued that a crucial *second stage* of the Anthropocene was the "great acceleration" in the buildup of greenhouse gases after World War II. These changes, enhanced by the threat of nuclear weapons, epitomized not only the transformation of the earth and its atmosphere but also the dangers to humans and to all life wrought by the Anthropocene. On a visit to the Max Planck Institute in 2012, Crutzen stated, "I'm starting to think the strongest signal, one of them, is just nuclear explosions—the test cases of atomic material.... Now I'm more in favor of declaring the nuclear tests as the real start of the Anthropocene." This concept was also proposed in a *Bulletin of the Atomic Scientists* article of 2015 titled "Can Nuclear Weapons Fallout Mark the Beginning of the Anthropocene Epoch?"[22]

Scholars have also invented new and different names for the

eras in which human actions have changed the nonhuman world: the Homogenocene, Plantationocene, Chthulucene, Gynocene, and Capitalocene.[23] What are these concepts, how do they differ from each other, and can they be reconciled?

The term *Homogenocene,* introduced in 1999 and elaborated by Charles Mann in his 2011 book *1493: Uncovering the World That Columbus Created,* refers to the ever increasing homogenization of biological life as both humans and invasive species have taken over more and more areas of the world. It is derived from the Greek words *homo* (same), *geno* (kind), *kainos* (new), and *cene* (period). In 1607, Jamestown Colony in Virginia, as portrayed by Mann, represented the advent of the Homogenocene in North America. Mann writes that Jamestown "was a brushfire in a planetary ecological conflagration."[24]

Yet another term, the *Plantationocene,* signifies the use of exploited working-class and slave labor on large plantations in colonized areas of the New World from the sixteenth century onward. It continues today in the form of mono-cropping, agribusiness, and factory farms for vegetable, meat, and milk production. All of these forms of industrialized agriculture reduce biodiversity. Donna Haraway invented the term *Chthuluocene,* which means tentacular thinking. She derives its name from "a spider, *Pimoa cthulhu,* who lives under stumps in the redwood forests of Sonoma and Mendocino Counties, near where I live in North Central California." Nets, networks, and kinships are keys to understanding the Chthuluocene. "Making kin" in the Chthulucene means to collect the trash created by humans in the Anthropocene and to shred, chip, and layer it into new forms of compost for the earth in the present and future.[25]

Still another term, although not as well developed, is the *Gynocene* (and its opposite, the *Patriarchalocene*). In the Patriarchalocene

INTRODUCTION

*Figure I.9. Donna Haraway (b. 1944)*

environmental deterioration is placed at the doorstep of patriarchal dominance against women and the earth. It can be reversed by feminist and ecofeminist actions (the Gynocene) along with indigenous movements that will help to restore ecological, gender, and human diversity.[26] Among the most significant of the alternatives to the Anthropocene is the Capitalocene, discussed below.

All these concepts and starting points for the transformation of the globe by human influence have been inspired by Crutzen and Stoermer's one-page article published at the start of the new millennium—the year the world began to take notice of the fate confronting humankind in the twenty-first century.

### The Climate of History

In his article "The Climate of History: Four Theses," published in *Critical Inquiry* in 2009, Dipesh Chakrabarty, then at the University of Chicago and subsequently a professor at the Institute for Advanced

INTRODUCTION

*Figure I.10. Dipesh Chakrabarty (b. 1948)*

Study at Princeton, detailed several issues pertaining to humans as a geological force in the Anthropocene. He analyzed the ways in which humans influence and are transformed by interactions with the global system. In so doing, he put forward four theses concerning the relationships between climate change and human history.[27]

First among Chakrabarty's theses is the collapse of the distinction between natural history and human history. Past interpretations of history have assumed that the human world is the primary focus, with the environment and nature being a passive backdrop. But with the rise of environmental history in all fields of history during the 1970s, scholars have increasingly paid attention to the role of humans as actors who can alter nature. In the Anthropocene, humans are likewise agents who are changing the chemistry and future of the earth, in a way similar to the climate variations that obliterated the dinosaurs.

Chakrabarty's second thesis is that with humanity as a new geological force, the very character of the history of modernity and globalization is also being transformed. For example, eighteenth-century theories that eulogized human freedom and equality must now also consider the changes in energy sources from wood to coal as an integral part of Enlightenment history. The new emphasis on the Anthropocene that sees humans as a geological force could challenge the very idea of human freedom. Will humans continue to find liberation from oppression or will their new home become a "planet of slums" with food and energy crises that challenge "basic happiness and dignity?"[28]

The third thesis concerns the idea of global histories of capital in conversation with the species history of humans. Here the important point is that the concepts of capitalist expansion and globalization cannot adequately account for all the ways in which humans are an earth-altering force. A more comprehensive look into history as "deep history" will clarify the ways humans have altered the planet over the past several thousand years. The intelligence and creative abilities that have allowed human beings to become dominant forces on the planet cannot be reduced to capitalism or socialism alone. We are now using those very technologies, such as the burning of fossil fuels, to threaten life on the planet. Nevertheless, the history of the human species as anthropos must be combined with the history of capitalism because industrialization in the nineteenth century could not have occurred without capital investments and cheap labor.

Chakrabarty's fourth thesis deals with ways to probe the limits of human understanding. Because the distinctions between human history and natural history have merged into a single history of humans in interaction with the environment, we need new interpretations of ourselves as geological agents. The crisis, therefore, cannot be reduced to one of capitalism alone. Climate change may not have come about intentionally, but we must accept that we as a species have produced it.

INTRODUCTION

## Capitalism and the Capitalocene

A strong rival to the term *Anthropocene* is the *Capitalocene*. In her 2014 book *This Changes Everything: Capitalism vs. the Climate* and her 2015 "Radical Guide to the Anthropocene," Naomi Klein asserts that capitalism is the culprit, not humans or human nature. The areas in which climate change has been most disruptive to human livelihoods are those places that also experience racial injustice. They are indeed human "sacrifice zones." Americans, she points out, consume five hundred times more energy than do—or can—Ethiopians. Poverty-stricken areas of the United States are also regions of inequality. We need to find new, non-exploitative ways of utilizing nature and promote them through grassroots movements so that in the future the planet will be a livable place.[29]

Another writer who blames capitalism is Canadian Ian Angus. Angus, who identifies himself as a Socialist, is the editor of "Climate and Capitalism," a Socialist history project. He has written "Confront-

*Figure I.11. Naomi Klein (b. 1970)*

INTRODUCTION

Figure I.12. Ian Angus (b. 1945)

ing the Climate Change Crisis: An Ecosocialist Perspective" (2008), *The Global Fight for Climate Justice* (2009), and *Facing the Anthropocene: Capitalism and the Crisis of the Earth System*. In 2015 he asked the question: "Does anthropocene science blame all humanity?" He points out that the wealthy countries have accounted for 80 percent of all carbon dioxide emissions since 1751, while the poorest countries account for less than 1 percent. He argues that the updated graphs by the IGBP highlight global inequities. Climate change is not a direct result of population growth, as neo-Malthusians would assert, but a capitalist-caused problem.[30]

The perspectives of native peoples offer a critical counterpart to capitalism's impact on nature. People should not be exploiters of the natural world; they should be one with it. Native peoples can provide examples of future ethics and actions. Eduardo Viveiros de Castro of the Federal University of Rio de Janeiro advocates this approach. He shows how indigenous cultures related to and influenced nature in his article "Exchanging Perspectives." He notes that most Amerindian

INTRODUCTION

*Figure I.13. Eduardo Viveiros de Castro (b. 1951)*

cultures did not differentiate between humans and other animals, an ontology that he calls "perspectivist animism." The attributes of humans and animals were combined, so that humans were not that different from nonhuman beings. Over the course of time, animals lost their human qualities, but they are really humans disguised in animal forms. The worlds that other animals inhabit and perceive are different from our human worlds. Only shamans can bridge the gap and communicate with animal worlds.[31] In confronting the impacts of capitalism and the Anthropocene on nature, the perspectives of native peoples around the world can therefore be critical. Additional research on these perspectives for the Anthropocene is needed.

The case for capitalism as the culprit and the Capitalocene as the major contender against the term *Anthropocene* has been made profoundly by Jason W. Moore of Binghamton University, New York. He has written insightfully about the tensions between the Anthropocene and the Capitalocene in articles and also in his 2016 edited book, *Anthropocene or Capitalocene?* Moore maintains that "we are living, not in the Anthropocene, but the Capitalocene." Moreover, this new era

## INTRODUCTION

*Figure I.14. Jason W. Moore (b. 1971)*

began not with the steam engine in 1874 but in 1450 with the rise of capitalist civilization, global conquest, and the relations of power, knowledge, and capital. During the "long sixteenth century," New World explorations and the global conquest of the planet by Europeans established first European hegemony and then preindustrial capitalism around the world. The solution? "Shut down a coal plant and slow global warming. But shut down the relations that made the coal plant and stop global warming for good."[32]

### Resolving the Differences

How can the differences among the above approaches to the Anthropocene be resolved? An important aspect of the anthropos, as captured by the term *Anthropocene,* is that humankind (which the term represents) applies to all people of the world—of every race, ethnicity, and skin color; rich or poor; male or female. All humans have the capacity to think and learn, speak and write, listen and hear. That ability means that everyone is capable of discovering truths through logic and mathematics and inventing machines through technology. People everywhere, therefore, are able to control nature through science and technology, and people of all genders, sexualities, races, and ethnicities have the capacity to do science and mathematics and to construct machines and digital devices. The ability to think empowers every marginalized person around the globe with the ability to learn science and technology. But within these capacities, there is a range of abilities and a vast difference in educational opportunities. Education is critical to understanding environmental issues and to participating in solving the problems of climate change.

Capitalism, on the other hand, as reflected in the term *Capitalocene,* organizes economic and social relations and creates profits and power, inequalities and oppression. Owners of businesses and factories deploy technologies, hire and pay workers, and reap profits generated by the enterprise. Risk is always involved in start-ups, and some succeed while others go bankrupt. Different countries have higher or lower percentages of successful businesses; particular businesses need greater or lesser quantities of natural resources, and more or fewer low-cost laborers.

Humans as anthropos therefore have created the science and technologies that have resulted in today's greenhouse gases, global warming, melting ice, and sea level rise. But different social and economic inputs in different countries—humans as *capitalos*—have produced different results. Some countries benefit and grow rich; others are exploited and become poor. Some countries produce more greenhouse gases and contribute more to global warming than others. The distribution throughout the globe is uneven.

Building on the foregoing discussion, we may argue that the concept of the Anthropocene subsumes the Capitalocene, but the Anthropocene is nevertheless implemented by the Capitalocene. Thus humans of all sexes, genders, races, and ethnicities have the brain power to do science and mathematics and to construct machines and digital devices (although abilities and education vary greatly), including technologies that can help make a better future. Colonial expansion initiated in Europe at the end of the fifteenth century created the preconditions for industrial capitalism. Human discoveries of both the power of steam (technology) and thermodynamics (science) were necessary conditions for industrial capitalist expansion. The technology of the steam engine supported the movement of manufacturing into cities and consolidated capital and labor into corporations. The Capitalocene thus constituted the socioeconomic relations between capital and labor that allowed Anthropocenic discoveries and anthropogenic effects to be disseminated throughout the world. Capitalist profit was based on cheap nature (fossil fuels) and cheap labor (slaves, immigrants, and the poor).

The process by which all this takes place is dialectical. Nature and humanity exist in interactive, process-oriented relationships. Nature and humanity are both dynamic and shape each other, such that nature⇔humanity. History is based not on simple cause-and-effect relations but on give-and-take processes. Similarly, achieving a viable future will be a back-and-forth process, characterized sometimes by progress and at other times by slowdowns or regressions.

Humans (as anthropos) have been able to dominate nature because of the mechanistic science (mathematics and experimentation) that developed during the Scientific Revolution of the seventeenth century *and* through technologies such as mining and the steam engine. But it was by means of capitalism and capitalist relations that entrepreneurs gained control over the fossil capital (coal, oil, gas) and labor (race/gender) that allowed for global expansion.

As a result of the burning of fossil fuels, greenhouse gases began

## Table 0.1

| Name | Function |
|---|---|
| *Anthropos* | Brains, science, technologies, inventions, solutions, modeling |
| *Capitalos* | Power, economic and social organization, capital plus cheap labor |
| *Politicos* | Policies, politics, democracy, negotiation, discussion |
| *Natura* | Living and nonliving beings, ecological relations |
| *Dialectica* | Process, interaction, humans and nature |

to fill the atmosphere. The resulting smoke and steam became symbols of human dominance. Both alarm and ambivalence were expressed through art and literature about the effects of smoke on humans and nature. As illustrated by graphic images and descriptions and as described in the following chapters, the effects were and are not evenly distributed among peoples and places.

## Synthesis

We may synthesize the foregoing ideas through the creation and use of five terms and the ideas they represent (see table 0.1).

These five concepts are tools with which to analyze the historical, scientific, technological, and political events that took place following the advent of the Anthropocene in the late eighteenth century and its effects on succeeding centuries and the future. Further research could focus more specifically on non-Western cultures and the origins and meanings of the Anthropocene for places such as Asia, Africa, Australia, Central and Latin America, the Arctic, and Antarctic.[33]

# 1
# History

Here I explore the period from the Enlightenment of the eighteenth century through the mid-twenty-first century, an era dubbed the Anthropocene by Paul Crutzen and Eugene Stoermer.[1] They argue that beginning with the introduction of James Watt's steam engine in 1784, humans have dramatically altered the earth's climate. This period also introduced a full-scale industrial, capitalist society culminating in what we are now experiencing, a "second death of nature."[2] This means that the human species itself, as digital mechanist, data analyst, and environmental manipulator par excellence, has potentially set up the preconditions for its own extinction. Do we need a new narrative, ethic, and worldview, along with new sciences and technologies that together can help to offset potential disaster in the Anthropocene? I conclude that the answer is clearly yes. But first, how did we get into this situation historically and how can we get out of it?

## Advent of the Anthropocene

What were the developments in science and technology that led to the concept of the Anthropocene as a new geological and ecological era beginning in the late eighteenth century? The Enlightenment that followed the so-called Scientific Revolution in the sixteenth and seventeenth centuries was a period of great optimism. The advances in science that stemmed from Sir Isaac Newton's *Principia Mathematica*

(1687) led to a belief that humans had the ability to understand and control nature. The ideas of Jean-Jacques Rousseau, Adam Smith, Voltaire, David Hume, Immanuel Kant, and other *philosophes* promoted scientific understanding, religious freedom, political independence, and the equality of all humans. New compilations of human knowledge of the world appeared in the form of Rousseau's *Discourse on Inequality* (1754) and *The Social Contract* (1762), Adam Smith's *Wealth of Nations* (1776), and Denis Diderot and Jean d'Alembert's *Encyclopédie* (1751–72). Academies, salons, and journals discussed and dispersed new knowledge of the natural world and its applications.[3]

Especially important for our purposes were the scientific discoveries that resulted in the burning of enormous quantities of fossil fuels and the pumping of $CO_2$ and other greenhouse gases into the atmosphere, ultimately leading to the Age of the Anthropocene. In 1754, Joseph Black discovered that by heating limestone (calcium carbonate) and treating it with acids, he could produce a gas that he called "fixed air" ($CO_2$). This substance would not support a flame or life itself. He further showed that this fixed air was produced by animals. In 1762 he introduced the concept of latent heat—the idea that a substance such as water will remain at the same temperature until the entire volume vaporizes, a fact critical to the workings of the steam engine.[4] Antoine Lavoisier in 1778 coined the term *oxygen*, "an eminently respirable part of the air," and discovered that it would support combustion.[5] Most important for our discussion of the Anthropocene, however, was the process by which James Watt's steam engine converted fossil fuels into greenhouse gases.

## James Watt's Steam Engine

The task of moving objects other than by human (or animal) labor is an age-old problem. The five simple machines of the Greeks—the

# HISTORY

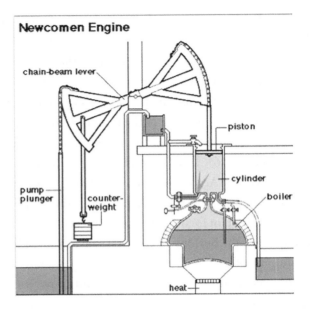

*Figure 1.1. Newcomen engine*

lever, the pulley, the wheel, the inclined plane, and the wedge—were force-maximizing devices, but they needed to be powered by human or animal labor. In the Middle Ages, watermills used the force of gravity in the form of falling water to move objects, while windmills used moving air to accomplish similar tasks.

Around 1712, preacher and ironmonger Thomas Newcomen (1664–1729), building on the work of Denis Papin, Thomas Savery, and others, invented an operating steam engine. By burning wood or coal in a furnace, water in a boiler was converted into steam that would expand to push a piston in a cylinder upward. Then, by condensing the steam with a shot of cold water, a vacuum was created on the top of the steam piston and external atmospheric pressure pushed it down, pulling the rocker arm upward. The rising and falling of the piston could then move the rocker arm, which pushed,

*Figure 1.2. James Watt (1736–1819), by John Partridge after Sir William Beechey, 1806*

pulled, raised, or lowered external objects without the use of human or animal labor.[6]

The Newcomen engine was immediately put to use all over England and greatly increased productivity. It was especially employed to pump water out of coal mines. The problem was that when the steam was cooled by injecting a shot of cold water to create the vacuum, it also cooled the cylinder. The cylinder then had to be reheated so that more steam could be created for the next motion of the piston, thereby wasting fuel.

In 1769, James Watt, working on a small model of the Newcomen engine at Glasgow University, began to improve its efficiency. He had been schooled by his mother at home; taking after his grandfather, he soon showed great ability at mathematics and engineering design. After studying instrument making, he was put in charge of the instrument collection at Glasgow University.

At Glasgow, while working on a model of the Newcomen engine, he discovered that adding an exterior unit in which conden-

# HISTORY

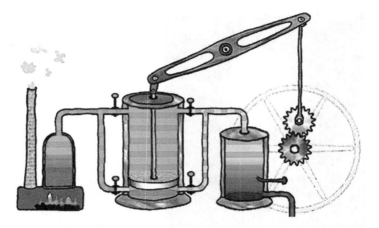

*Figure 1.3. James Watt steam engine*

sation could take place meant that he did not have to waste fuel by repeatedly heating and cooling the same cylinder. Steam, created in the boiler, expanded into the cylinder where the piston was located. The expanding steam then pushed the top of the piston downward. Then a separate condenser filled with cold water sprayed water into the steam above the piston, reducing the air pressure and drawing the piston upward. With stopcocks placed both above and below the piston, the steam and low pressure could act alternately in a double action that enormously increased efficiency.

In 1784, Watt and his partner Matthew Boulton patented a diagram of a double-acting steam engine and that diagram was used to construct steam engines all over England. It was soon adapted beyond raising coal from mines, and steam-driven textile mills, steamboats, and steam trains were developed.[7]

## The Steam Engine and the Second Law of Thermodynamics

In the mid-nineteenth century, physicists Sadi Carnot, Benoît Paul Émile Clapeyron, and Rudolf Clausius tackled the difficult problem of improving the amount of mechanical work obtained from the James Watt steam engine. In the process, they discovered that there can never be a perfect engine with no loss of heat—a discovery that by the 1850s became the basis for the second law of thermodynamics.

Sadi Carnot was the brilliant son of mathematician and French Revolutionary Army leader Lazare Carnot, who taught Sadi and his brother Hippolyte mathematics, physics, music, and languages.[8] Sadi entered the Paris École Polytechnique when he was only sixteen years old and graduated at the age of eighteen. He served in the French Army Corps of Engineers until 1819 when, after taking a leave of absence, he began to think about how to improve James Watt's steam engine. Was there a way to create a model engine that could operate at 100 percent efficiency, he asked, by converting all the heat that went

*Figure 1.4. Sadi Carnot (1796–1832)*

into it into useful work? At that time the most sophisticated steam engine known had an efficiency of only 3 percent. What would an ideal steam engine look like and how would it work?

In 1824, at the age of twenty-eight, Carnot published a short book, *Reflections on the Motive Power of Fire*. In it, he showed that the efficiency of the steam engine depends only on the temperatures of the two heat reservoirs in the cylinder and condenser and that an ideal engine would be frictionless and independent of the fluid used.[9]

In eloquent prose he began:

> Everyone knows that heat can produce motion. That it possesses vast motive-power no one can doubt, in these days when the steam engine is everywhere so well known.
>
> To heat also are due the vast movements which take place on the earth. It causes the agitations of the atmosphere, the ascension of clouds, the fall of rain and of meteors, the currents of water which channel the surface of the globe, and of which man has thus far employed but a small portion. Even earthquakes and volcanic eruptions are the result of heat....
>
> From this immense reservoir we may draw the moving force necessary for our purposes. Nature, in providing us with combustibles on all sides, has given us the power to produce, at all times and in all places, heat and the impelling power which is the result of it. To develop this power, to appropriate it to our uses, is the object of heat-engines.
>
> The study of these engines is of the greatest interest, their importance is enormous, their use is continually increasing, and they seem destined to produce a great revolution in the civilized world.[10]

Carnot titled his book *Reflections on the Motive Power of Fire*. But what did he mean by motive power? And what did he mean by fire? The "motive power" was the amount of work (that is, force acting through a distance) accomplished by the upward and downward motion of the piston. Carnot used "fire" to refer to the heat source of the steam engine—the wood or coal that was burned in a furnace. He also worked within the framework of belief, prevalent at the time, that heat was a substance called caloric rather than the motion of molecules, as was later determined.[11]

Carnot developed what was later named the Carnot cycle (the term itself was first used in 1887).[12] For Carnot, the ideal steam engine consisted of a cylinder, a piston, a working substance such as water (that could be converted to steam), a heat source (wood or coal), and a sink (or reservoir for the cooled-down steam). The Carnot cycle referred to the process whereby a gas expands as it absorbs the heat produced by the boiler and then condenses as it is cooled forcefully by a blast of water. The amount of "motive power," or work that a steam engine could produce, was critical to its efficiency. Carnot showed that the level of efficiency depended only on the differences between the two temperatures and not on the particular gas used for the expansion and compression of the piston.[13]

Carnot died of cholera in 1832, and his book on the "motive power of fire" went virtually unnoticed in his lifetime. Remarkably, however, it was revived and reformulated two years after his death by Émile Clapeyron who, like Carnot, had studied at the École Polytechnique in Paris. Clapeyron designed steam engines and later brought his designs to England in an effort to engage a manufacturer who could produce a more efficient model. In 1834, he wrote his first memoir (which bore a similar title to that of Carnot's work): "Memoir on the

*Figure 1.5. Benoît Paul Émile Clapeyron (1799–1864)*

Motive Power of Heat." And, like Carnot, he worked within the caloric theory of heat. At the outset, he noted the significance of Carnot's work and emphasized the importance of restating Carnot's theory in mathematical terms.[14]

It has long been known, Clapeyron began, both that heat (caloric) can produce "motive power" (work) and that in turn "motive power" can produce heat. In steam engines, in the process of producing work, the heat produced by combustion (that is, by burning coal in a furnace) is always accompanied by heat collected in the condenser at a lower temperature. "A certain quantity of caloric *always* passes from one body at a given temperature to another body at a lower temperature." In other words, it was known that all the heat could not be transformed into work, but it was necessary to prove it mathematically so that values such as heat and temperature could be quantified and the amount of work predicted.[15]

Clapeyron then drew a trapezoidal figure that represented the Carnot cycle, or the mechanical force developed by the change in the quantity of heat (caloric) transferred from the higher temperature vessel (the boiler) to a lower temperature vessel (the condenser). This change, as Carnot had shown, was independent of the gas or vapor used. Clapeyron concluded that the "caloric, passing from one body to another maintained at a smaller temperature, can give rise to the production of a certain quantity of mechanical action." Using an extensive set of differential equations, he then proceeded to derive the mathematical relationships that existed when the vapor (steam) at a higher temperature is changed to a lower temperature to produce mechanical work.[16]

Eric Mendoza (1919–2007), formerly a professor at Manchester University, UK, who introduced and reprinted the papers of Carnot, Clapeyron, and Clausius in 1960, argues that Clapeyron not only contributed to the articulation of the second law of thermodynamics, he also unambiguously stated what became the first law of thermodynamics: "It follows," Clapeyron wrote, "that a quantity of mechanical action, and a quantity of heat which can pass from a hot body to a cold body, are quantities of the same nature, and that it is possible to replace the one by the other; in the same manner as in mechanics a body which is able to fall from a certain height and a mass moving with a certain velocity are quantities of the same order which can be transformed one into the other by physical means."[17] Clapeyron thus advanced the concepts that became the first and second laws of thermodynamics while still working within the caloric theory that heat was a substance.

Then in 1850, a twenty-eight-year-old scientist named Rudolf

*Figure 1.6. Rudolf Clausius (1822–88)*

Clausius followed with a paper, "On the Motive Force of Heat, and on the Laws Which Can Be Deduced from It for the Theory of Heat." Clausius, who proved to be a brilliant mathematician, was born in what is now Poland and educated in a school taught by his father. He went to the University of Berlin, where he studied mathematics, physics, and history before getting his doctorate in 1847 from the University of Halle in Germany. He then taught in Berlin, Zurich, and Bonn as a professor.

Clausius's "On the Motive Power of Heat" was published in the German journal Poggendorff's *Annals of Physics* (*Annalen der Physik*). Here he abandoned the term *caloric* and used Clapeyron's concept of heat. He noted that only the first part of what Carnot had written was correct, namely, that "the equivalent of the work done by heat is found in the mere transfer of heat from a hotter to a colder body." He then stated (without naming it as such) what became known as the second law of thermodynamics: "A transfer of heat from a hotter to a colder body always occurs in those cases in which work is done by heat, and in which also the condition is fulfilled that the working substance is in the same state at the end as at the beginning of the operation."[18]

In 1856 Clausius published a second paper, "On a Modified Form of the Second Fundamental Theorem in the Mechanical Theory of Heat." Here he restated what became known as the second law as follows: "Heat can never pass from a colder to a warmer body without some other change, connected therewith, occurring at the same time." In other words, for heat to be transferred from a cold to a hot body, work has to be expended.[19]

In 1865, he published another work, "On the Mechanical Theory of Heat: With Its Applications to the Steam Engine." In this paper he coined the term *entropy* for the loss of energy available to perform work and used the term *energy* instead of *motive force*. He also stated the "two fundamental theorems of the mechanical theory of heat (1) the energy of the universe is constant and (2) the entropy of the universe tends to a maximum."[20] That is, in closed systems isolated from their surroundings, the energy available for work (moving an object through space) is always decreasing. Entropy, the energy unavailable for work, is always increasing. In practice, a steam engine would have to be fully insulated from the environment in a container that does not allow heat to enter or escape from within its entire structure: a closed, isolated system.[21] As later clarified by others:

> The example of a heat engine illustrates one of the many ways in which the second law of thermodynamics can be applied. One way to generalize the example is to consider the heat engine and its heat reservoir as parts of an isolated (or closed) system—i.e., one that does not exchange heat or work with its surroundings. For example, the heat engine and reservoir could be encased in a rigid container with insulating walls. In this case, the second law of thermodynamics (in the sim-

plified form presented here) says that no matter what process takes place inside the container, its entropy must increase or remain the same in the limit of a reversible process.[22]

## Thermodynamics as a Field of Science

The naming of thermodynamics and its development as a distinctive field took place in the early 1850s through the work of William Thomson (Lord Kelvin), and William Rankine. Thomson, like James Watt, did much of his work at Glasgow University on the banks of the River Kelvin. As a result of his discoveries in thermodynamics, he was honored with the title Baron Kelvin in 1892 and then, as the first scientist to serve in the House of Lords, became Lord Kelvin. The absolute scale of temperature (beginning at 0° Kelvin, or absolute zero), in which he stated that the lowest temperature possible is −273.15°C (or −459.67°F), is named after him.[23]

In 1851, Kelvin wrote "On the Dynamical Theory of Heat," in which (as had Clausius) he abandoned the term *caloric*, favoring in-

Figure 1.7. *William Thomson (Lord Kelvin) (1824–1907)*

stead the emerging idea that heat was a motion of material particles. He stated that considering that "heat is not a substance, but a dynamical form of mechanical effect, we perceive that there must be an equivalence between mechanical work and heat, as between cause and effect." He continued, "Heat is not a substance, but a state of motion." He also stated the second law of thermodynamics, in somewhat different terms: "It is impossible, by means of inanimate material agency, to derive mechanical effect from any portion of matter by cooling it below the temperature of the coldest of the surrounding objects."[24]

Kelvin's 1851 paper referred to the work of James Prescott Joule (1818–1889), who had rejected the caloric theory in a paper published in the *Philosophical Magazine*, "On Changes of Temperature Produced by the Rarefaction and Condensation of Air," in 1845.[25] Here Joule likewise discarded the caloric theory and proposed instead that "heat is regarded as a state of motion among the constituent particles of bodies." He described an experiment he had performed that used a falling weight to heat water by rotating a paddle wheel inside an insulated barrel. Joule wrote, "It is easy to understand how the mechanical force expended in the condensation of air may be communicated to these particles so as to increase the rapidity of their motion, and thus may produce the phenomenon of increase of temperature." Joule was prescient not only in rejecting the concept of caloric, but also in his idea that heat was the motion of particles. While on his honeymoon in 1847, he is said to have met Kelvin by chance and the two went to a waterfall in Chamonix in the Alps of southeastern France. There, with very sensitive thermometers, they attempted to measure the temperature difference between the water at the top and bottom of the waterfall.[26] The theory was that the water at the bottom would be warmer than at the top because the kinetic energy of the falling water was converted to heat when it came to rest at the bottom.[27]

In 1852, Kelvin published, "On the Universal Tendency in Nature to the Dissipation of Mechanical Energy," which ultimately led to the concept of the "heat death" of the universe.[28] Kelvin did not use the term *heat death,* however, and did not extend the concept beyond the earth itself.[29] Nevertheless, he reached three conclusions concerning the implications of the dissipation of energy for the future of the earth: "(1) There is at present in the material world a universal tendency to the dissipation of mechanical energy. (2) Any restoration of mechanical energy, without more than an equivalent of dissipation, is impossible in inanimate material processes, and is probably never effected by means of organized matter, either endowed with vegetable life or subject to the will of an animated creature. (3) *Within a finite period of time past, the earth must have been, and within a finite period to come the earth must again be, unfit for the habitation of man as at present constituted,* unless operations have been, or are to be performed, which are impossible under the laws to which the known operations going on at present in the material world are subject."[30]

Here he set out the idea that the earth is vulnerable, not because of human action (as in the concept of the Anthropocene), but because of the relentless and irreversible action of the second law of "thermodynamics"—a term he introduced in 1854. He stated: "Thermodynamics is the subject of the relation to forces acting between contiguous parts of bodies, and the relation of heat to electrical agency."[31]

When generalized beyond what Kelvin stated about the earth to the universe at large, a "heat death" implies that if the universe itself can be considered an isolated system, then its entropy also increases with time and ultimately there is no temperature differential left to perform work. As later defined by others: "The implication is that the universe must ultimately suffer a 'heat death' as its entropy

# HISTORY

*Figure 1.8. William Rankine (1820–72)*

progressively increases toward a maximum value and all parts come into thermal equilibrium at a uniform temperature. After that point, no further changes involving the conversion of heat into useful work would be possible. In general, the equilibrium state for an isolated system is precisely that state of maximum entropy. (This is equivalent to an alternate definition for the term *entropy* as a measure of the disorder of a system, such that a completely random dispersion of elements corresponds to maximum entropy, or minimum information.)"[32] But whether a "heat death" in an ever-expanding universe is still a theoretical possibility, given recent developments in astronomy and physics, is now debatable.[33]

In 1859, William Rankine continued the development of the field when he wrote the first textbook on thermodynamics, *A Manual of the Steam Engine and Other Prime Movers*.[34] In a chapter titled, "Of the Two Laws of Thermodynamics," he explained both the first and second laws of thermodynamics. The fact that his book's title referenced the steam engine underscored the significance of James Watt's invention in initiating the discoveries that led to the laws of thermodynamics.

By the late nineteenth century, what became known as classical (or near-equilibrium) thermodynamics that applied to closed, isolated systems, such as steam engines, refrigerators, and chemical systems, had been fully formulated owing to the work of Carnot, Clapeyron, Clausius, Joule, Kelvin, Rankine, and others. With the subsequent development of statistical thermodynamics, including Ludwig Boltzmann's probability equation for the entropy of an ideal gas as $S = k.\log W$, which was engraved on his tombstone, much of the field of thermodynamics seemed to be complete.[35]

### Far-from-Equilibrium Thermodynamics

In the late twentieth century, however, Belgian physicist Ilya Prigogine (1917–2003) (see chapter 5 below) in his Nobel Prize–winning research challenged the implications of the second law of thermodynamics with his concept of far-from-equilibrium thermodynamics and dissipative structures. In his Nobel Prize lecture of 1977, he stated: "It is the main thesis of this lecture that we are only at the beginning of a new development of theoretical chemistry and physics in which thermodynamic concepts will play an even more basic role."[36]

Prigogine argued that classical thermodynamics holds in systems that are in equilibrium or near equilibrium, such as pendulum clocks, steam engines, and solar systems. These are stable systems in which small changes within the system lead to adjustments and adapta-

*Figures 1.9 and 1.10. Ludwig Boltzmann's (1844–1906) tombstone in Vienna and his equation for entropy (ch 2, note 35)*

tions. They are described mathematically by the great seventeenth- and eighteenth-century mathematical advances in calculus and linear differential equations. But what happens when the input is so large that a system cannot adjust? In these far-from-equilibrium systems, nonlinear relationships take over. In such cases small inputs can produce new and unexpected effects.

Prigogine's far-from-equilibrium thermodynamics allows for the possibility that higher levels of organization can spontaneously emerge out of disorder when a system breaks down. His approach applies to social and ecological systems, which are open rather than closed, and helps to account for biological and social evolution. In the biological realm, when old structures break down, small inputs can (but do not necessarily) lead to positive feedbacks that may produce new enzymes or new cellular structures. In social terms, revolutionary changes can take place. On a large scale, a social or economic revolution can occur in which a society regroups around a different social or economic form, such as the change from gathering-hunting to horticulture, or from a feudal society to a preindustrial capitalist society. In the field of science, a revolutionary change could entail a paradigm shift toward new explanatory theories, such as the change from a geocentric Ptolemaic cosmos to a heliocentric Copernican universe.[37]

Together, the two laws of thermodynamics, as developed in the nineteenth century, summarized the relationships between heat-energy and work (moving an object through space) and helped to explain the reasons that the steam engine's efficiency was limited. The first law stated that the energy of the universe is a constant and is changed only in form (for example, from mechanical to electrical to chemical to biological). The second law stated that the total energy in the universe available for doing work is always decreasing; conversely, the entropy is always increasing. While the second law applied to closed systems in near equilibrium (such as the steam engine and the refrigerator), far-from-equilibrium thermodynamics, developed in the late twentieth century, held that under certain circumstances, open systems could reorganize themselves into new forms (from the cellular to the societal level).

The second law of thermodynamics was of immense consequence in the historical period following the James Watt steam engine of the 1780s. The optimism of the eighteenth-century Enlightenment faded, exposing new limits to reality. But although what people could actually accomplish on the earth seemed severely compromised, the steam engine nevertheless took off. It became the basis for the steamboat, the train, the factory, and the age of industrialization, spewing carbon dioxide from the burning of fossil fuels into the atmosphere. Ultimately, with the internal combustion engine in automobiles and airplanes, followed by diesel-powered machines, more and more $CO_2$ was pumped into the air and oceans—resulting in global warming. Although I have focused this chapter on James Watt's steam engine and the history of thermodynamics, additional research is called for on the history and consequences of steam power in other countries around the world.

The Age of the Anthropocene, in which humans are capable of causing a new "death of nature" on the planet, is now our twenty-first-century nightmare. The art, literature, and poetry discussed in the next chapters underscore the potentially irrevocable impacts set off by the steam engine on humanity and on the earth—and what must be done to reverse them.

# 2
# Art

Artists and photographers who have engaged with global warming believe that the visual arts are an essential part of creating the large-scale public awareness and understanding of climate change that can bring about substantive policy change. Diverse academics have investigated this view in recent years. The effects of art and photography on climate change can be significant. Not only in galleries and museums but almost everywhere in media, words are ceding space to images. Throughout newspapers, magazines, and institutional publications, issues are increasingly framed and perceived through images. Art, like literature, can reveal major changes to the landscape as global warming progresses. In this chapter, I show how artists challenge the standard human/environment narrative, in which humans are both privileged over other species and separate from nature. Indeed, artists can change the way we think about the meaning of progress.

## Steam-Powered Art

Steam power fueled much of the art and literature of the nineteenth century. The steam engine triggered the movement of manufacturing into cities and the consolidation of capital and labor into corporations. It also generated the transportation revolution (from horses and mules to steam trains and steamboats), the market revolution (from cooperation to competition and profits), and an ecological revolution that produced air pollution, greenhouse gases, and ecological change from the burning of fossil fuels.

The human domination of nature initiated by the Scientific Revolution was enhanced by new technologies, especially the steam engine, and the mining of fossil fuels—coal, oil, and gas. In the process of industrial expansion, capital became concentrated among factory owners and elites, while labor was supplied by working-class people earning low wages. Labor thus became a tool for capitalist expansion. Greenhouse gases were manifested in the atmosphere as smoky air and plumes of black steam. As expressed in art and literature, the smoke and steam from steamboats, trains, and factories become symbols of human dominance. At the same time, however, these creative works transmitted ambivalence over what the new technology did to both humans and nature. Graphic images and descriptions flooded the art and literature of the nineteenth century. Today, however, we can interpret them from the perspectives of both human progress and environmental decline.

### Railways and Industries

The first passenger railway was introduced in England in 1825 by George Stephenson, who built the Stockton and Darlington Railway. It was operated by a steam-driven engine called the Rocket. This was soon followed by the Manchester-Liverpool railway in 1830. The effects of steam engine smoke on human life were immediately visible and of enormous concern. A letter to the *Leeds Intelligencer* of January 13, 1831, stated: "On the very line of this railway, I have built a comfortable house; it enjoys a pleasing view of the country. Now judge, my friend, of my mortification, whilst I am sitting comfortably at breakfast with my family, enjoying the purity of the summer air, in a moment my dwelling, once consecrated to peace and retirement,

Figure 2.1. Steam engine crossing the landscape

is filled with dense smoke of foetid gas; my homely, though cleanly, table covered with dirt; and the features of my wife and family almost obscured by a polluted atmosphere. Nothing is heard but the clanking iron, the blasphemous song, or the appalling curses of the directors of these infernal machines."[1]

The rural landscape was contaminated by tails of "smoke curling across the countryside." Those who lived near the London and Birmingham railroad in 1825 asked whether their cows' grazing fields would be damaged and if the quality of their sheep's wool would be ruined by the black smoke. Would the fox runs that they needed in order to walk through their fields be destroyed? And would their chickens still be able to lay eggs?[2]

Stationary steam engines liberated the textile industry from dependence on falling water to operate mills. They added power when creeks ran dry in the summers and enabled the mills to be located in

# ART

*Figure 2.2. Stationary steam engine built by W. Bradley, Gooder Lane Ironworks, Brighouse, in the 1880s, Stott Park Bobbin Mill*

cities. Sheep's wool was spun into thread, wound onto bobbins, and woven on looms to produce textiles. The textiles were then dyed, cut, and sewn into clothing. The Stott Park Mill in Cumbria, England, for example, built in 1835 to produce wooden bobbins for mills in Lancashire and Yorkshire, added a steam engine in the 1880s.

### Early Paintings of Steam Engines

The English artist Joseph Turner painted several works of steam-driven boats and trains asserting themselves into the landscape.

Turner's 1838 *Fighting Temeraire Being Tugged to Her Last Berth* exhibits the "raw mechanical energy of the flaming and smoking tug."[3] And in Turner's *Rain, Steam, and Speed: The Great Western Railway* of 1844, "a train rushes across a bridge and is bearing down on a hare that is running over the washed-brown bed of a railway track."[4] The observer, powerfully experiencing the unrelenting speed of the train,

*Figure 2.3. Joseph Turner (1775–1851),*
Self-portrait, *ca. 1799*

*Figure 2.4. Joseph Turner,* Fighting Temeraire
Being Tugged to Her Last Berth, *1838*

ART

*Figure 2.5. Claude Monet (1840–1926), Arrival of the Normandy Train, 1877*

is led to ask: can the hare escape? Can anyone escape? Can humanity escape? The painting expresses an appreciation for the speed of the new coal train and appreciation for the new technology of transportation, but it also raises troubling questions. What is the threat that humans pose for the earth? What should we fear most? The awe of the wild or its annihilation?

France's railway station La Gare St. Lazare, which opened in the 1870s, likewise stimulated paintings of trains arriving in stations. In 1870, Édouard Manet (1832–83) painted *The Railroad Station in Sceaux* in the suburbs of Paris. The landscape is shrouded in white snow, gray clouds, and bleak fields.[5] In 1877 Claude Monet painted the *Arrival of the Normandy Train*, a mixture of impressionism and realism.[6] The black engine arrives against a background of blurred white and bluish smoke and clouds. It signifies the dominance of human-built tech-

## ART

*Figure 2.6. A train barrels down the tracks in the Lumières' 1896 film*

nologies over the small black human figures in the station. The question becomes: how can such technology be controlled? What is the fate of humanity as smoke and pollution produced by the monster engines take over the world?

These images reveal the progress of steam transportation and industrialization beyond England and their continuing power to inspire artists to capture their implications for humankind on canvas. One of the very first movies ever made, in 1895, *The Arrival of a Train at La Ciotat Station,* depicted a railway station in southeastern France. When the film, produced by Auguste and Louis Lumière, was first shown in January 1896, it is said to have caused panic in the theater, with people running for the exits.

### Art in the United States

By the early nineteenth century, steam-driven boats appeared throughout the United States on rivers and canals, soon to be followed by the construction of railroads and increasing numbers of steam engines. Steam power also made it possible for manufacturing plants and factories to move into cities. In 1811, the first stationary steam engine was built for the Middletown Woolen Manufacturing Company of Connecticut. By 1838, 317 steam engines were operating textile mills in New England. Railways were introduced as a means of transporting textiles and other manufactured goods; the effects of steam and its black smoke on the countryside soon became manifest.

On waterways, steamboats replaced canal boats pulled on towpaths by mules. Locks raised and lowered steamboats along the waterways and canals that linked factories to markets. The earliest canals along which steamboats traveled included the Middlesex Canal, which connected Lowell, Massachusetts, with Boston in 1803; the Erie Canal, which traversed upstate New York in 1825; the Blackstone Canal, which linked Providence, Rhode Island, to Worcester, Massachusetts, in 1828; and the Great Lakes to Ohio and Mississippi Canal, completed in the 1830s.

Numerous nineteenth- and twentieth-century paintings illustrate the impact of the steam engine on the American landscape. Images of steamboats and factories showed black smoke pouring from chimneys. The idea that the steam engine could adversely transform nature began to take hold in American art in the late nineteenth century. In Andrew Melrose's 1867 *Westward the Star of Empire Takes Its Way*, a train dominated by a bright headlight explodes out of the forest. Deer attempting to flee stand motionless on the tracks, transfixed by the light. As historian William Cronon points out, a strong diagonal line divides the painting. Wilderness, the sublime, is on the right; on the other side is the pastoral. All the trees have been cut down in order to construct the farmhouse in the clearing. The wilderness is now violated by the steam train. The deer are fleeing before the light, but where can they go? Their wilderness home has been destroyed and a human home created in the transformed landscape. They have totally lost their place in nature as the *Star of Empire* makes its way westward.[7]

Many nineteenth-century painters thought of wilderness as a positive thing, yet they also celebrated the progress of America. Their paintings, therefore, tell ambivalent stories: excitement about the up-

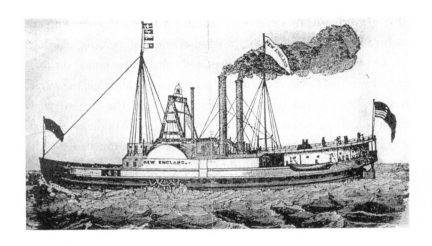

*Figures 2.7 and 2.8. Nineteenth-century images of steamboats and factories*

ART

*Figure 2.9. Andrew Melrose's,* Westward the Star of Empire Takes Its Way, *1867, depicts an area near Council Bluffs, Iowa*

ward trajectory of American progress and despair over the banishing of nature. The artists are reflecting, enjoying, and appreciating the idea of empire and progress even as they lament it.

The steam-powered transportation system converted the U.S. into one vast, unified market, with sectional specialization in staples and manufactured products. These sections were the South, the Northeast, the Middle Atlantic region, and the West. Linking East with the West was the cross-continental railway system, with the Union Pacific and Central Pacific Railroads meeting at Promontory Point in Utah in 1869. Built largely by Chinese labor, the transportation network converted the U.S. into a single cross-country market.

The celebrations at the joining of the two railroads lauded the achievement of conquering the country's vast terrain, lowering its hills, and straightening its curves. The Reverend Dr. Dwinell's sermon at the 1869 dedication of the transcontinental railroad quoted Isaiah 40:4: "Prepare ye the way of the Lord, make straight in the desert

ART

*Figure 2.10. John Gast,* American Progress, *1872*

a highway before our God. Every valley shall be exalted, and every mountain and hill shall be made low and the crooked shall be made straight and the rough places plain."

John Gast unambiguously depicts the cross-country narrative in his 1872 painting *American Progress*. On the left, toward the west, is Nature active, alive, wild, dark, and savage—filled, as William Bradford would have put it, with "wild beasts and wild men." Buffalo, wolves, and elk flee in dark disorder, accompanied by Indians with horses and travois. On the right, to the east, is Nature ordered, civilized, and tamed. No longer to be feared or sexually assaulted, she floats angelically through the air in flowing white robes, emblazoned with the star of empire. She carries telegraph wires in her left hand, symbols of the highest level of communication—language borne through the air, the word or logos from above. The image of the domination of logic or pure form is repeated in the book grasped in her right hand, touching the coiled telegraph wires. She represents the city, the civil, the civic order of government—the highest order of nature. She is pure Platonic form impressed on female matter, transforming and ordering all beneath her. Most important, however, it is American men who have prepared her way. They have dispelled the darkness, fought the Indian, killed the bear and buffalo. Covered wagons bearing westward pioneers, gold rush prospectors, and the Pony Express precede her. Farmers plowing the soil next to their fenced fields and rude cabins have settled and tamed the land. Stagecoaches and steam trains follow, bringing waves of additional settlers. At the far right is the Atlantic civilization, where ships bearing the arts of the Old World arrive in the New World. The painting itself is a lived progressive narrative. Its east to west movement is a story of ascent and conquest. Yet the spewing black smoke heralds the coming of the Age of the Anthropocene.[8]

In 1931, John Kane painted *Monongahela Valley*, which also celebrated industrialization. His painting focuses almost entirely on industry, and conveys the triumph of progress. There is no intimation of anything disturbing or problematic about inserting industry into the landscape. Black fumes pour out of coal-burning factories and the coal-burning engines on the steam-driven barges. But mostly the steam is white and benign. In the painting's foreground is a fence outlining the field. The railroad station in the center brings passengers into the Monongahela Valley, while the river flows downstream toward Pittsburgh, scene of iron and coal industries. Anthracite coal extracted from the Pennsylvania landscape is the primary source of fuel for the steam engines. Grouped along the road are the factories

ART

*Figure 2.11. John Kane (1860–1934),* The Monongahela River Valley, Pennsylvania, *1931*

and houses of the industrial workers. On the right, a steam-driven freight train moves across the landscape.[9]

The painting reveals several sources of steam power, the transformative force driving industrialization: the steam barge, the factories, and the steam engines that propel the train. Along the painting's diagonal are docks where the raw materials brought by the trains and the manufactured products from the factories are loaded onto barges and shipped out of town. Storage containers for goods and barrels for lumps of coal characterize a scene of burgeoning industry. The transportation revolution combines river travel with train travel and the two come together to make the market revolution possible.

But, despite the artist's celebration of progress, an element of criticism appears, whether intentional or not. There are blemishes on this landscape. The hills in the background are denuded of trees, and the straight-line grid pattern of the fields is imposed on the terrain. In this painting eulogizing late nineteenth- and early twentieth-century progress lies the beginnings of an environmental critique, as revealed in air and water pollution and a deformed landscape. Still, the primary story here is one of progress: smoke is good and is necessary to economic development. Today we might read these symbols very differently than did people living a hundred years ago. For us, the painting has a sense of declension. From an environmentalist perspective, we might view these elements of advancement as ugly and deformed. Yet, for our nineteenth-century counterparts, they were monuments of positive change and transformation.

### Working on the Railroad

If the steam engine is the hallmark of the Anthropocene, fossil fuels its source of energy, and steam art one of its most potent manifestations, then those who labored to create and implement the power of steam are its strength: from the scientists, engineers, and technologists who designed and improved the engine to the manufacturers who produced multiple models over time to the men and women who were employed or enslaved as laborers "working on the railroad." It is beyond the scope of this book to detail the various production systems and improvements, but the types of laborers who built and maintained the engines and tracks deserve some attention. Here art and imagery can further illuminate this dimension of the Anthropocene.

Prior to the Civil War, black slaves worked on all the major railroads in the eastern United States and continued as "negro workers" after the war ended. In the West, Chinese labor was used to build

ART

*Figure 2.12. Trackwomen at the Baltimore & Ohio Railroad Company, 1943*

the railroads that moved from west to east to join those being constructed across the Great Plains and Rockies. Women likewise worked to maintain the railroads and in some cases were employed as engine drivers, especially during World War II. Several illustrations reveal the onerous work and endless hours that went into railroad construction and maintenance. Such images expose the realities of labor in the Anthropocene, here appropriately interpreted as the age of man—or, in even more negative terms, the Patriarchalocene, the Androcene, or the Slavocene. In recent years, however, women and blacks have become locomotive engineers and executives in the railroad industry.[10]

*Figure 2.13.* African American Railway Workers: *railroads bought slaves or leased them from their owners, usually for clearing, grading, and laying tracks. Enslaved workers frequently appear in annual reports as line-item expenses, referred to variously as "hands," "colored hands," "Negro hires," "Negro property," or "slaves."*

*Figure 2.14. Female engineer, Llangollen Railway, Wales, 2009*

# ART

## Art of the Future

If we can move beyond the Age of the Anthropocene into an Age of Sustainability by the end of the twenty-first century, what will the art of the future look like? If the world moves away from fossil fuels and toward solar power and renewable energy sources, as many predict (see the epilogue to this book), images such as steam engines, steamboats, and black smoke could serve merely as reminders of a past age filled with greenhouse gases, air pollution, and unhealthy living. Autonomous cars, electric underground trains, bicycles, and work-at-home modes of life powered by the sun could become the basis for new art forms.

Here are a few technological changes, as described by engineer Tony Thorne, that could become the basis for the art of the future:[11]

> Autonomous cars: In 2018 the first self-driving cars appeared for public use. Around 2020, the complete industry will start to be disrupted. You don't want to own a car anymore. You will call a car with your phone, it will show up at your location and drive you to your destination. You will not need to park it, you only pay for the driven distance and you can be productive while driving. Our children may never get a driver's license and may never own a car.
>
> Autonomous vehicles will change our cities, because we will need 90–95% fewer cars. We could even transform former parking spaces into parks.
>
> Electric cars will become mainstream during the 2020's. Cities will be less noisy because all new cars will run on electricity. Electricity will become incredibly cheap and clean: Solar production has been on an exponential curve for 30 years, and soon we will see the burgeoning impact.

## ART

Last year, more solar energy was installed worldwide than fossil fuel energy. Energy companies are desperately trying to limit access to the grid to prevent competition from home solar installations, but that can't last. Technology will take care of that strategy.

With cheap electricity will come cheap and abundant water. Desalination of salt water now only needs 2kwh per cubic meter (@ 0.25 cents). We do not have scarce water in most places, we only have scarce drinking water. Imagine what will be possible if anyone can have as much clean water as s/he wants, for nearly no cost.

Artists have in fact proposed images and representations of a new era that critiques the age of fossil fuels and climate change. Cutting-edge works include those of Olafur Eliasson. At a San Francisco Museum of Modern Art exhibition in the fall of 2007, visitors could don a gray blanket and enter a room kept at exactly 10°F to see Eliasson's "ice car," a BMW hydrogen-powered racing car covered in a thick coat of ice. Eliasson hopes his "ice car" will spur thinking on the relation between car design and climate change. Like many artists addressing the issue of climate change, Eliasson uses art to create environmental awareness, engaging and prompting the public to acknowledge responsibility and foster social change. He hopes his work will inspire more responsible public behavior. Eliasson states, "What I find so interesting in this research on movement and environmentally sustainable energy is the fact that it enhances our sense of responsibility regarding how we navigate as individuals in our shared, complex and polyphonic world."[12]

The Cape Farewell Project leads Arctic expeditions for artists, scientists, and journalists in hopes of increasing environmental aware-

*Figure 2.15. Olafur Eliasson's* Your Mobile Expectations: BMW H₂R Project, *2007*

ness and engaging the public and schools in more fruitful debates about climate change. Project founder David Buckland believes in the power of art to help bring about policy change. "One salient image, sculpture or event," says Buckland, "can speak louder than volumes of scientific data and engage the public's imagination in an immediate way."[13]

Similarly, journalist Alex Morrison said of an exhibit traveling the world from 2007 through 2008, *Envisioning Change,* which chronicles the effects of climate change on diverse global locales such as the polar regions, the Andes, and the Himalayas: "The beautiful, thought-provoking, and sometimes shocking images engage viewers on an emotional level that can't be achieved through words alone."[14] The goal is to increase awareness of the effects of climate change on the world's coldest regions and to inspire changes in behavior that can slow it down.

Increasingly, with most individuals living in cities, fewer people enjoy childhoods or even summers in the wilderness. In increasingly crowded cities, art is one of the best ways to educate large numbers of Americans about the rich heritage of wilderness aesthetics that played such a large role in fueling past and present environmental movements, from the parks movements of the early twentieth century to the sweeping legislative changes of the 1970s to the climate change movement today. Rapid changes taking place in parks and forests are a vital spark for the current effort. To name just one poignant example, Glacier National Park is facing the declassification of a large number of its glaciers.

Images could play a formative role in changing personal behavior and public policy in the future. Viewing art could help to promote individual and collective action. Climate change may be directly experienced only as a few hot days in summer or connected to a powerful hurricane, but that beginning awareness, aided by the powerful influence of art, could be used to stimulate behavioral changes over longer time frames. Thus, viewing spectacular canvasses could indeed inspire social change. More examples of art beyond those I have introduced in this chapter should be studied; I hope in particular that scholars will broaden the scope beyond Europe and the United States to include other countries and continents and that future works will elaborate the roles of race, class, and gender.

# 3
# Literature

A rich history of literature exists that is highly relevant to the Age of the Anthropocene. This history includes poets such as William Wordsworth, Walt Whitman, Robert Frost, Gary Snyder, and Robert Hass; and writers such as Charles Dickens, Nathaniel Hawthorne, John Steinbeck, Ralph Waldo Emerson, Henry David Thoreau, Aldo Leopold, John McPhee, Barbara Kingsolver, and Annie Dillard. These authors all reflected the immense changes that took place with the speeding up of life resulting from the steam engine and the devastation of the environment from the impacts of coal and smoke—hallmarks of climate change and the Anthropocene.

Many prominent figures in the environmental humanities were deeply influenced by the changes to life wrought by industrialization. For example, Aldo Leopold was a naturalist, ecologist, and farmer as well as a "green" philosopher who grew up on a bluff in Iowa overlooking the railroad and courted his wife to be, Estella, while the two held hands and walked the tracks of the Santa Fe Railroad. Annie Dillard grew up in the wilds of Pennsylvania's rivers and forests and grieved over the impact of industrialization on her later home in the Roanoke Valley of Virginia. And writers such as John McPhee and Barbara Kingsolver find their inspiration in the remains of American "wilderness" while grieving over the impact of climate change on the land and endangered species.[1]

## LITERATURE

*Figure 3.1. William Wordsworth (1770–1850), portrait by Samuel Crosthwaite, 1844*

In this chapter I discuss writers in England and the United States who responded to the early impacts of coal and the steam engine as well as current authors who discuss the meaning of the Anthropocene for peoples around the world today and in the future. Further research and writing on this topic is needed as we go forward into the twenty-first century.

### British Literature

As early as 1814, William Wordsworth complained of the negative effects of human actions on nature that would come to characterize the Anthropocene epoch. He viewed smoke as a major threat to the environment: it brought suffocation to all of life and had long-lasting effects on nature. In "The Excursion" he complained of human intrusion into the natural world:

> Hiding the face of earth for leagues—and there,
> Where not a Habitation stood before,

> The Abodes of men irregularly massed
> Like trees in forests—spread through spacious tracts,
> O'er which the smoke of unremitting fires
> Hangs permanent, and plentiful as wreaths
> Of vapour glittering in the morning sun.
> And, wheresoe'er the Traveller turns his steps,
> He sees the barren wilderness erased,
> Or disappearing ...[2]

The subsequent presence of steamboats and steam engines likewise blighted nature, but not human vision. Smoke could mar the beauty of the natural world, but the mind could still retain its prophetic insights. In "Steamboats, Viaducts and Railways," published in 1833, Wordsworth wrote:

> Nor shall your presence, howsoe'er it mar
> The loveliness of Nature, prove a bar
> To the Mind's gaining that prophetic sense
> Of futur change, that point of vision, whence
> May be discovered what in soul ye are.[3]

The Kendal and Windermere railroad line that opened in 1847 threatened Wordsworth's beloved Lake District home. Wordsworth lent his voice to the preservation of nature, opposing the technological takeover by steam-powered industries even before the nearby railroad was completed. In a poem published in the *London Morning Post* in 1844, "On the Projected Kendal and Windermere Railway," Wordsworth asked, "Is there no nook of English ground secure from rash assault?" He insisted that Nature should speak out: "Speak, passing winds; ye torrents, with your strong / And constant voice, protest

LITERATURE

*Figure 3.2. The* William Wordsworth

against the wrong."[4] In another poem written that year denouncing the proposed railway line, he wrote:

> Hear ye that whistle? As her long-linked Train
> Swept onwards, did the vision cross your view? . . .
> Mountains, and Vales, and Floods, I call on you
> To share the passion of a just disdain.[5]

In a twist of irony, in 1952, the Crew Company (which was naming engines after famous British literary figures), named one of its locomotives *William Wordsworth*.

Charles Dickens was another British author who made the steam engine a major actor in his writings. In his novel *Dombey and Son* (written between 1846 and 1848), Dickens employs vivid language to describe the impact of the train on land and life. The construction of the London and Birmingham Railway in Stagg's Gardens (a neighborhood in Camden Town, London), he declared, had an effect akin to an earthquake that changed the entire landscape into a grotesque monster.[6] The railroad, like "the first shock of a great earthquake . . . rent the whole neighborhood to its center." Traces of its course were visible on every side. Houses were knocked down; streets broken through and stopped; deep pits and trenches dug in the ground; enormous heaps of earth and clay thrown up; buildings that were undermined and shaking, propped by great beams of wood. Here, a chaos of carts, overthrown and jumbled together, lay topsy-turvy at the bottom of a steep unnatural hill; there, confused treasures of iron soaked and rusted in something that had accidentally become a pond."[7]

The neighborhood's transformation by the new railroad made it virtually unrecognizable. Bridges went nowhere, roads were impassable, chimneys stood above the ruins of destroyed dwellings, and ashes, bricks, and planks were strewn in horrific disarray.[8] "In short, the yet unfinished and unopened Railroad was in progress; and, from the very core of all this dire disorder, trailed smoothly away, upon its mighty course of civilization and improvement."[9] Indeed, the speed with which life passes, plunging its denizens into an unknown future is so great that its meaning becomes impenetrable. No longer is there a way to savor or understand the people and landscapes through which a horse and carriage used to carry its occupants. Death approached faster and more inexorably than ever before. On the train, meaning becomes incomprehensible, "as in the track of the remorseless monster, Death!"[10]

In his 1854 novel *Hard Times,* Dickens laments the impact of smoke on working-class towns. Indeed, in the novel the town itself, appropriately named Coketown, is the epitome of the irretrievable damage the industrial revolution has done to the countryside.

LITERATURE

*Figure 3.3. Charles Dickens (1812–70), portrait by William Powell Firth, 1859*

A SUNNY midsummer day. There was such a thing sometimes, even in Coketown. Seen from a distance in such weather, Coketown lay shrouded in a haze of its own, which appeared impervious to the sun's rays. You only knew the town was there, because you knew there could have been no such sulky blotch upon the prospect without a town. A blur of soot and smoke, now confusedly tending this way, now that way, now aspiring to the vault of Heaven, now murkily creeping along the earth, as the wind rose and fell, or changed its quarter: a dense formless jumble, with sheets of cross light in it, that showed nothing but masses of darkness:—Coketown in the distance was suggestive of itself, though not a brick of it could be seen.[11]

In the hot summer days, exhausted by shoveling coal into steam engines, the stokers were greeted after work by streets and yards dusty with soot, the smell of hot oil, and the relentless hum of the mill shafts and wheels. Unable to find relief from brutal working days, they struggled on into the evening twilight and hot humid nights.[12] Such were the effects of the industrial revolution in its home country of England. The effects were felt even more profoundly in its overseas expansion into New England. Here trains, steam engines, and factories began to dominate a landscape once pristine.

### American Literature

American novelist Nathaniel Hawthorne, like Dickens, saw stark contrasts between life on a speeding train versus the past stability of the homestead. Indeed, the train's speed symbolizes the acceleration of life in the Anthropocene. The feelings of loss and despair over nature and life that the steam engine invoked in the nineteenth century are similar to those experienced today by many people living in the Anthropocene.

In *The House of the Seven Gables* (1851), Hawthorne portrayed the train as representing both freedom and fear, progress and dread, excitement and horror. The character Clifford feels the freedom of rushing into the unknown and the excitement of new and different places and scenes. But home, on the other hand, represents the safe, the secure, and the comfortable. At home, in the house of the seven gables, one has the choice of viewing people out of the arched front window, but on the train people are everywhere and unavoidable. There is no privacy or escape. One must continually endure encounters with others. Constantly traveling, confronting new places, and dealing with unknown situations is too taxing and challenging. Returning home is essential to life itself.[13]

In *The Celestial Railroad* (1843), Hawthorne writes of a railroad journey from the City of Destruction to the Celestial City in the com-

LITERATURE

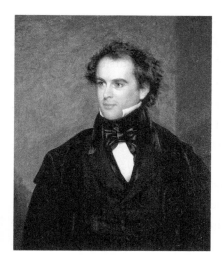

*Figure 3.4. Nathaniel Hawthorne (1804–64), portrait by Charles Osgood, 1841*

pany of a director of the railroad corporation, Mr. Smooth-it-away. To get to the future on the celestial train, the stagecoach (symbol of the past) first has to cross a bridge covering a highly disagreeable quagmire made of "twenty thousand cartloads" of ancient "outdated" texts, consisting of "books of morality, volumes of French philosophy and German rationalism; tracts, sermons, and essays of modern clergymen; extracts from Plato, Confucius, and various Hindu sages together with a few ingenious commentaries upon texts of Scripture—all of which by some scientific process, have been converted into a mass-like granite."[14]

The train's passengers are all ladies of elegance and gentlemen of reputation who engage in politics and business. The train's engine, however, resembles a mechanical demon that might take its passengers to a netherworld rather than the Celestial City. The chief engineer is "a personage almost enveloped in smoke and flame, which ... appeared to gush from his own mouth and stomach as well as from the engine's brazen abdomen."[15]

LITERATURE

Once on the train and swept up in its thunderbolt voyage, the travelers can see the voyagers of the past—foot travelers with staffs and back burdens groaning and stumbling in pursuit of their goals, fanning the smoke and scalding steam away from their faces as they trudge on their way. Then comes a "tremendous shriek, careering along the valley as if a thousand devils had burst their lungs to utter it, but which proved to be merely the whistle of the engine on arriving at a stopping-place."[16]

Toward the end of the voyage the travelers encounter a new form of steam. "A steam ferry boat, the last improvement on this important route, lay at the river side, puffing, snorting, and emitting all those other disagreeable utterances which betoken the departure to be immediate." The passengers hurry on board, terrified that the boat will sink or explode, pale from breathing in the steam and frightened by the appearance of the ugly steersman. The voyager himself rushes to edge of the boat, intending to escape overboard, but the paddlewheels churn and turn, spitting cold freezing spray overboard, chilling him so deeply that it wakens him from his dream, shivering and quaking. So much for the modernization wrought by the steam engine. So much for the advantages of rapid travel through space and time into a celestial future. The Celestial Railroad is a mere illusion, an idealistic vision, an Edenic dream that cannot become reality without tremendous consequences for humanity and the earth.[17]

In Leo Marx's *The Machine in the Garden* (1967), Americans experience the disruptions of what would later be called the Anthropocene. All across the country, trains, boats, and factories oppose barns, fields, and pastures. The age of agriculture differs from the age of industry, just as the Holocene contrasts with the Anthropocene. The disruption of steam technology opposes the peacefulness of the pasture; its power and force uproot the stability of the countryside. The garden is a place of beauty, a site in which food is produced, a place

## LITERATURE

*Figure 3.5. Ralph Waldo Emerson (1803–82)*

out of time. The machine is the rush of time, taking over the farm and transforming life itself. The question that arises is whether and how both can survive. Can both be accommodated? Is there a dialectic, a give-and-take, between permanence and change, between past and future?[18] The American dream is filled with discordance.

In his essay "The Young American" (1844), Ralph Waldo Emerson praised the railroad and the advantages it brought to Americans. "An unlooked consequence of the railroad," he wrote, "is the increased acquaintance it has given the American people with the boundless resources of their own soil." "Railroad iron," he said, "is a magician's rod, in its power to evoke the sleeping energies of the land and water." In 1871, he took a trip on the transcontinental railroad two years after it had been completed. Traveling all the way to California, he visited Yosemite, where he met John Muir. The railroad had indeed opened new vistas and exciting new opportunities for Emerson.[19]

Henry David Thoreau's *Walden; or, Life in the Woods* (1854), by contrast, exemplifies the dialectic and disruption wrought by the train. In 1845 Thoreau retreated to Walden Pond just outside of Con-

LITERATURE

Figure 3.6. Henry David Thoreau (1817–62), daguerreotype by Benjamin D. Maxham, 1856

cord, Massachusetts, where he lived for two years and two months, leaving on September 6, 1847. The train ran along edge of Walden Pond. Its whistle pierced the air like a hawk and he could hear the rattle on the rails from his cabin. But even as it disturbed his solitude, the train became a symbol of the market that brought new and different resources and commodities from around the world.

Like Hawthorne and Emerson, Thoreau was influenced by the railroad at a time when the machine wreaked havoc on the country's pastoral ideal. The railroad represents the next step upward in a continual narrative of American progress from colonial times to the era of industrialization to today's Anthropocene epoch. A new type of landscape is created that can be totally dominated by humans, one in which people can move rapidly from place to place. For Thoreau, the iron horse plants its foot firmly on the ground and its neighing penetrates the silence of his Walden cabin. The "machine in the garden" underscores technology's continuing impact on the American landscape and the ongoing loss of the American wilderness.

The immense power and speed of the railroad, as a new form of

LITERATURE

*Figure 3.7. Walden train station*

transportation unlike any other, made an enormous impact on all observers. Mark Twain, in *Roughing It* (1872), conveys shock and awe concerning the power of the train careening across the continent and lighting up the wilds. He writes, "About 4:20 pm, we rolled out of the station at Omaha. A couple of hours out, dinner was announced, an event for those of us who had yet to experience what it is to eat in one of Pullman's hotels on wheels. As we washed the good things down with bumpers of sparkling champagne, our train with its great glaring Polyphemous eye, lighting up the long distance of the prairie, rushed into the night and the Wild."

Walt Whitman, on the other hand, eulogized the "fierce-throated beauty" of the engine in his "To a Locomotive in Winter," first published in 1886 in "Two Rivulets" and subsequently in 1900 in *Leaves of Grass*. In the lines "Thy black cylindric body, golden brass and silvery steel" and "Thy ponderous side-bars, parallel and connecting rods,

*Figure 3.8. Mark Twain (1835–1910), ca. 1907*

*Figure 3.9. Walt Whitman (1819–92), photograph by Mathew Brady*

gyrating, shuttling at thy sides," he praised the engine's manufactured materials—iron and steel—and the new era of industrialization that would come to characterize the Anthropocene. Like Mark Twain, he was awed by "thy great protruding head-light fix'd in front." He called the steam engine a new "type of the modern—emblem of motion and power—pulse of the continent." The train was a law unto itself, an earthquake, a body belching murky clouds of smoke as it plunged forth into the wintry gale. In March 2003, Whitman's poem was set to music by Greg Bartholomew and performed a cappella by the choral group Seattle Pro Musica.[20] The Anthropocene was now a musical.

The implications of the early years of the Anthropocene and life's accelerated pace spring forward in American poetry. In "The Railway Train," first published in 1896 after her death, Emily Dickinson likened the train to a horse that lapped up the miles, licked up the valleys, and fed itself at tanks, all the while neighing before stopping "omnipotent/at its own stable door." The train is alive—personified with needs, emotions, and power. It reflects the ambivalence many Americans felt at the takeover of the landscape by a being more powerful than themselves and yet helpless without human upkeep.[21]

> I like to see it lap the miles,
> And lick the valleys up,
> And stop to feed itself at tanks;
> And then, prodigious, step
>
> Around a pile of mountains,
> And, supercilious, peer
> In shanties by the sides of roads;
> And then a quarry pare
>
> To fit its sides, and crawl between,
> Complaining all the while

In horrid, hooting stanza;
Then chase itself down the hill

And neigh like Boanerges;
Then, punctual as a star,
Stop—docile and omnipotent—
At its own stable door.

Robert Frost (1874–1963) in his poem "A Passing Glimpse," published in *West-Running Brook* (1928), lamented the experiences and visions missed by riding on the whizzing train. He wished to stop, get off, and look at the flowers by the wayside. Their names were lost as the train rushed madly past. The poem symbolizes the limits of the human ability to comprehend a world overcome by technology.[22] Like the other writers discussed here, he uses the imagery of the railroad and the immense speed of the steam engine to gain insight into

*Figure 3.10. Emily Dickinson (1830–86), daguerreotype, 1846–47*

humanity's diminishing relationship with nature. They lament the inability of people to understand life as the full impact of the Anthropocene begins to bear down upon them.

### Recent Literature

Recent writers continue to use the steam train to emphasize how life abruptly changed in the twentieth century in ways that anticipated the full impact of the Anthropocene. In his poem "A Stone Garden," originally published in 1959 in *Riprap and Cold Mountain Poems*, poet Gary Snyder (b. 1930) writes of waking up from dreaming on a train to confront a future filled with despair.

> I thought I heard an axe chop in the woods
> It broke the dream; and woke up dreaming on a train.
> It must have been a thousand years ago
> In some old mountain sawmill of Japan.
> A horde of excess poets and unwed girls
> And I that night prowled Tokyo like a bear
> Tracking the human future
> Of intelligence and despair.[23]

In *Pilgrim at Tinker Creek* (1974), Annie Dillard describes the changes that have taken place outside her home in the Roanoke Valley of Virginia in the Blue Ridge Mountains. Concerning the effect of the train on her peaceful, beautiful environment, she imagines the problem the manager of the Southern Railroad faces. He has to manufacture the engines that will pull the railroad cars up the steep grade between the cities of Lynchburg and Danville, Virginia. He finances the production of nine thousand engines at an enormous cost. "Each engine must be fashioned just so, every rivet and bolt secure, every

*Figure 3.11. Annie Dillard (b. 1945)*

wire twisted and wrapped, every needle on every indicator sensitive and accurate." The problem is that although each engine has an engineer to manage its speed and all the trains are sent out, there are no switch operators! Is the Anthropocene about to crash?

> The engines crash, collide, derail, jump, jam, burn.... At the end of the massacre you have three engines, which is what the run would support in the first place. There are few enough of them that they can stay out of each other's paths.
>
> You go to your board of directors and show them what you've done. And what are they going to say? You know what they're going to say: It's a hell of a way to run a railroad.
>
> Is it a better way to run a universe?[24]

For Dillard, trains and railroads have transformed not only nature but also her community, her life, and seemingly the universe itself.

Trains symbolize irrevocable change from the local to the global, from the peaceful past to the screeching future, from aimless wandering in fields of flowers to inexorable acceleration into a universe of iron and steel. With fire and fury, they drag us along into a world filled with smoke and soot, fueled by coal and oil, and on into a future called the Anthropocene.

In 2005, John McPhee (b. 1931) wrote a two-part article in the *New Yorker* titled "Coal Train"—a dramatic and profound description of the ways in which coal is transported by rail across the United States and its significance for climate change. McPhee rides the train with engineer Scott Davis and conductor Paul Fitzpatrick, who give him a firsthand lesson in how the initially empty five-engine, 7,433-foot train is run, controlled, and loaded. It travels westward from Marysville, Kansas, all the way to the Black Thunder Mine in the Powder River Basin between Montana and Wyoming—source of 40 percent of all the coal mined in the United States. From the Powder River, some thirty-five coal trains are constantly traveling back and forth to the Robert W. Scherer Power Plant, just outside of Macon, Georgia, the largest coal-fired electrical plant in the United States. Powder River coal became significant after the passage of the Clean Air Act of 1970 because it "was five times lower in sulfur than Appalachian coal ... [and] power plants were required to scrub sulfur out or burn low sulfur coal."[25]

In the second part of McPhee's article, a loaded coal train leaves the Powder River Basin for its journey east. McPhee relates that they met "coal trains, auto trains, rock trains, and grain trains," many of which were empty and returning whence they came for more cargo. Their train passed twenty miles of motionless trains queued up on the vast plain, waiting their turn to get onto the tracks. "Direct-current diesel-electric locomotives," writes McPhee, "are fine for hauling auto-trains, intermodal containers, and sugar beets, but alternating current is the better way to move the weight of coal.... A coal train is so heavy that it should be limited to a hundred cars if the locomotives are only on the front end."[26] The immense amount of pollution produced during these constant trips across the continent is a huge contribution to climate change and the acceleration of the Anthropocene.

Amitav Ghosh, in *The Great Derangement: Climate Change and the Unthinkable* (2016), argues that "global warming's resistance to the arts begins deep underground, where organic matter undergoes transformations that make it possible for us to devour the sun's rays." He notes that coal and oil are not substances that writers enjoy depicting. They are "viscous, foul smelling, repellant to all the senses." The two fossil fuels, however, are different in the ways they can arouse emotions. In coal extraction, miners are front and center, inspiring class solidarity and worker resistance, stimulating an expansion of worker rights in the late nineteenth century. By contrast, the depiction of oil extraction in Ghosh's novel *The Circle of Reason* (1986) occurs against a background of tall derricks fortified with wire fences in a very dehumanized setting. "Out of the sand, there suddenly arose the barbed wire fence of the Oiltown. From the other side of the fence, faces stared silently out—Filipino faces, Indian faces, Egyptian faces, even a few Ghaziri faces, a whole world of faces." Oil has changed the environments in which we live, but has "almost no presence in our imaginative lives, in art, music, dance, or literature." Ghosh nevertheless concludes that depictions of global warming though art and literature can help humanity to transcend the problems we face today and bring human beings closer to each other and to other beings on the planet.[27]

## Automobiles and Airplanes in the Anthropocene

The train differed from its successor transport, the automobile of the early twentieth century. Although destined to produce far more greenhouse gases in the Anthropocene, the development of the automobile was initially slow and ponderous, even as it represented a new form of capitalist enterprise that would soon vastly increase air pollution. Today, the transportation sector of the U.S. economy (cars, trucks, ships, trains, and airplanes) produces some 27 percent of the country's greenhouse gas emissions. The consumption of fossil fuels by cars and trucks can be reduced by hybrid electric and autonomous electric vehicles.[28]

But airplanes use thousands of barrels of fossil fuels each year. According to the U.S. Energy Information Administration, in the year 2017, U.S. airlines consumed 1,398 thousand barrels of jet fuel per day, by far the highest of any country in the world. China was second, with 388 thousand barrels per day. World jet fuel consumption was approximately 5.5 million barrels per day. According to research done by Erin Lo at the University of Pennsylvania, in 2016 jet fuel consumption by aircraft was responsible for 12 percent of all U.S. transportation greenhouse gas emissions and 3 percent of total U.S. greenhouse gas emissions across all sectors: "Jet fuel consumption accounts for a growing amount of greenhouse gas emissions and continues to rise with globalization. Due to its increasing share of greenhouse gas emissions, aviation has been thrust into the spotlight as an industry with the potential to abate carbon emissions. Despite collaborative intentions, many countries cannot come to an agreement for aviation standards and greenhouse gas reduction targets.... My research predicts jet fuel consumption levels in 2030 and 2050 will be 39.65% and 95.06% greater than 2013 jet fuel consumption levels, respectively."[29]

According to environmental journalist Prachi Patel, the use of biofuels could reduce the emissions of climate warming particles by up to 70 percent. Because biofuels have virtually no sulfur and carbon compounds, a bio-jet fuel blend could assist in the reduction of greenhouse gases.³⁰ Research, however, is only beginning on the use and implications of using biofuels.

If climate change is the hallmark of the Anthropocene, then novels can help to convey the urgent need for action. Barbara Kingsolver's novel *Flight Behavior* (2012) relates the impact of climate change on an iconic species, the monarch butterfly, in a community in Tennessee. The hero is a black Harvard-educated scientist who in Kingsolver's descriptions resembles Barack Obama. Kingsolver sent the novel to Michelle Obama. In February 2015, the Obama administration committed $3.2 million to saving the monarch, and the Monsanto chemical company pledged $4 million to the project.³¹

American poet Anne Waldman (b. 1945) sums up her concerns about the Anthropocene in a poem titled "Anthropocene Blues." Her poem speaks of the "tragedy of the Anthropocene" and laments the "new weathers" ahead. She writes of "climate grief" and asks whether we will fail to save our world. She concludes, "My love for you sings for you, world, I've got those Anthropocene Blues."³² Sam Solnick of the University of Liverpool asks what it means to write poetry in and about the Anthropocene. His 2016 book *Poetry and the Anthropocene* argues that we need poetry and the humanities to help us think in new ways to save our world.³³

Adam Trexler of Portland, Oregon, wrote *Anthropocene Fictions: The Novel in a Time of Climate Change*, in 2015. He holds that the novel is one of the best ways to search for meaning in this new era when climate change is an all-consuming fact of life. Novels help people grasp the changes occurring and what they mean to human existence as climate change becomes ever more significant in daily experience.³⁴

In 2016, the Association for the Study of Literature and the Environment (ASLE) called for contributions to a special issue of *Interdis-*

*ciplinary Studies in Literature and Environment,* "The Literature of the Anthropocene." It challenged its readers and writers to think deeply about how literature can respond to this new geological era and the ways in which writers and poets can help to resolve the problems facing all of humanity in the Anthropocene.[35]

## Women and Gender in the Anthropocene

Critical to literature in the Anthropocene is the place of women and gender. (Indeed, some have dubbed the new era the Androcene, the Patriarchalocene, or the Phallocene.)[36] Instead of the Anthropocene (or age of man), we should be entering the Gynocene, an age in which women can contribute policies and power to help resolve climate change. Melina Pereira Savi of the Federal University of Santa Catarina in Brazil and others argue that it is women who are the most affected by climate change, especially women in developing countries, on islands, and along the coasts. As oceans rise, coastal and riverine areas become uninhabitable, forcing thousands to move inland to higher ground where land is already densely settled and scarce. According to the United Nations, women are the most affected by environmental destruction and are the most impacted by ecological deprivation. Women, especially those in developing countries, must spend more time and energy carrying water from distant places, gathering firewood, and tending crops in depleted soils. Women should have the power to make changes that will affect their lives and the earth itself.[37]

Ecofeminism, which evaluates the history and cultural connections between women and nature, is particularly relevant to understanding the role of women in the new era. Melina Pereira Savi cites ecofeminist author Jane Bennett, who argues that the Anthropocene represents an opportunity to rethink the world in terms of the vitality

## LITERATURE

of living bodies. Nonhuman bodies and ecological forces are themselves agents of change that cannot be controlled by humans and must be taken into consideration. "How," Bennett asks, "would political responses to public problems change were we to take seriously the vitality of (nonhuman) bodies?" Nonhuman entities, such as storms, metals, food, and commodities, can not only "impede or block the will and designs of humans but also act as quasi agents or forces with trajectories, propensities, or tendencies of their own." Texts can guide humans to insights, meanings, and perceptions that can institute change.[38]

Savi concludes her analysis of gender and literature in the Anthropocene with the comment that "literature, like the humanities, is outpouring with works that warn, ponder on, and speculate what is happening and what might happen if we continue to overlook the practices that have led the world to enter (according to human parameters, of course) the Anthropocene Epoch." New ethics and behaviors can assist humans in creating more desirable ways to act to improve the future of the planet in the Anthropocene.[39]

Numerous individuals continue to write about literature in the Anthropocene. Lara Stevens, Peta Tait, and Denise Varney's edited book *Feminist Ecologies: Changing Environments in the Anthropocene* (2018) relates ecofeminist ideas to the Anthropocene. Alessandro Macilenti's *Characterising the Anthropocene: Ecological Degradation in Italian Twenty-First Century Literary Writing* (2018) examines Italian nature through chemical pollution, land transformation, and climates of fear in the future. Alice Major's *Welcome to the Anthropocene* contributes gripping poetry about the world that awaits us in the future of humankind on earth. All these books and poems are alerts for what is to come if humanity does not respond and act decisively to turn back the consequences of climate change and become sustainable.[40]

## LITERATURE

In conclusion, literature can help us understand the human dimensions of the Anthropocene in ways that can profoundly affect our interactions with nature and life on earth. From the speeding up of life on railroads zipping past flowers and pastures to the increased smog and pollution brought about by automobiles and airplanes, literature can help us to see that humans are no longer (and never really were) in control of nature. Nature is both autonomous and reactive and we must interact with it in ways that make possible a new partnership and new ways of dealing with climate change and greenhouse gases. Examples and analyses of the roles of literature in developing countries and the unequal causes and manifestations of climate change between the northern and southern hemispheres, beyond what I have attempted in this book, could help to spread the message worldwide. Increasingly, the poetry and novels of the twenty-first century will reflect the problems of the Anthropocene and offer ways to introduce changes that will save human and nonhuman lives and nature itself in the future.

# 4
# Religion

Established religions and individual spirituality form critical perspectives on the Anthropocene. As historian of religions Mary Evelyn Tucker points out, "Religion and ecology together now constitute a 'field' within academia and a 'force' in the larger society.... The field is now poised to be a key participant in dialogue involving the Anthropocene and environmental humanities."[1] In this chapter I examine the role of mainstream religions in mitigating the problems of climate change and the ways in which forms of spirituality can become moral guidelines for individual actions. Inasmuch as the Anthropocene concerns the means by which greenhouse gases are introduced into the atmosphere and climate change occurs, religion and spirituality together represent an antidote and a deterrent. Changes ranging from overseas work to assist other countries in providing sustainable energy sources to individual churches installing solar panels can make a difference. Renewable energy sources such as solar, wind, and hydropower can replace fossil fuels in a new Age of Sustainability. Locally produced panels and windmills manufactured by local labor offer alternatives to corporate carbon pollution. These goals can be accomplished within mainstream religious denominations as well as alternative forms of spirituality.

## Christianity and Western Culture

In his foundational essay of 1967, historian Lynn White Jr. identified Western Christianity as it emerged during the medieval period as providing a religious justification for the human domination of nature. As an alternative, he proposed St. Francis of Assisi as a patron saint for the stewardship of nature by humans. "Especially in its Western form," White asserted, "Christianity is the most anthropocentric religion the world has ever seen.... Christianity in absolute contrast to ancient paganism and Asia's religions ... not only established a dualism of man and nature but also insisted that it is God's will that man exploit nature for his proper ends."[2] Christianity challenged pagan ideas of an I/thou relationship of humans to nature in which all beings, both animate and inanimate, were alive. If humans destroyed a tree or animal or even mined the earth for "living" metals, nature might retaliate. During the sixteenth century, as the earth was mined for ores to support an expanding money-based economy and Christianity became the dominant religion, nature itself was devalued and seen as a source of capital. The "death of nature," as I argued in my 1980 book of that name, became associated with a God who was now a clockmaker, an engineer, and a mathematician.[3] Isaac Newton, in his 1687 *Mathematical Principles of Natural Philosophy,* synthesized the astronomical laws of Nicolaus Copernicus and Johannes Kepler with the terrestrial mechanics of Galileo Galilei, René Descartes, and Robert Boyle. The universe was no longer a living organism to be propitiated when harvested or mined, but a machine that was dead, inert, and exploitable.

In Western Europe, mainstream Christian traditions that emerged during the Renaissance and Reformation eras of the sixteenth and seventeenth centuries were carried around the globe during periods of colonial settlement by Europeans. Individual salvation and freedom from sin constituted hope for a positive afterlife. Although Protestantism and Catholicism did not intersect to any great extent with conservation until the late twentieth century, recent consciousness about sustainability has now infused mainstream religions.[4]

## Religion and Ecology

The mainstream narrative of Western culture that arose after the Renaissance and Reformation and extended from the Scientific Revolution to the present has been the re-creation of the entire earth as a managed Garden of Eden, with humans as its faithful stewards.[5] But global warming challenges both the possibilities for human management of the environment and the very predictability of environmental change itself. In searching for new ways of responding to change and alleviating the impacts of a warming world on humans, particularly the poor, and other life-forms, world religions have found themselves drawing on ancient traditions and spiritual roots. The centrality of religions in addressing climate change is no longer the exception but the rule.

In the United States and abroad, religious organizations are flocking to the cause of climate change. Indeed, it seems that every major denomination in the U.S. has a program to combat climate change, including Baha'i, Buddhist, Christian Ecumenical, Greek Orthodox, Unitarian, indigenous spiritual groups, Muslim, Jewish, Quaker, and more.[6] Some of the largest of these groups are interfaith organizations, bringing people together across dogmas and creeds to focus on concrete issues that can be agreed upon and acted upon, such as climate change.

Katharine Jefferts Schori is a former oceanographer and former presiding bishop of the Episcopal Church. The Episcopal Church held a "Healing Our Planet Earth" (HOPE) conference in Bellevue, Washington, in 2008 at which she spoke. The consensus was to reduce greenhouse gas emissions by 50 percent in ten years at every church, synagogue, and facility.[7]

Interfaith groups such as GreenFaith in New Jersey are making major inroads. GreenFaith's mission is to mobilize "religious institutions and people of diverse faith to strengthen their relationship with the sacred in nature and to take action for the earth." To this end the organization encourages conversion to renewable energies across the state. For instance, it is collaborating to place solar panels at twenty faith-based sites around New Jersey. The Interfaith Center on Corporate Responsibility has made "reversing global warming" one of its top priorities.[8]

The National Council of Churches, with approximately 45 million members of Protestant, Anglican, and Orthodox faiths, is one of the featured partners of StopGlobalWarming.org. And several large evangelical networks reach similar numbers of congregants. Just one of these groups, the Evangelical Climate Initiative, comprises over eighty-five U.S. evangelical leaders who have signed the statement "Climate Change: An Evangelical Call to Action."[9]

Academics likewise have established venues for studying and responding to climate change as it impacts ecology and humanity. The Forum on Religion and Ecology at Yale University highlights the important role that religions play in constructing moral frameworks for interacting with other people and the environment. It facilitates academic and engaged discourse on the intersection of religious studies, science, and environmental policy. At the Harvard Center for the Study of World Religions, Mary Evelyn Tucker and John Grim organized a series of ten conferences involving over eight hundred scholars and environmentalists. From these conferences they edited a nine-

RELIGION

*Figure 4.1. John Grim (b. 1946) and Mary Evelyn Tucker (b. 1949)*

volume book series, *Religions of the World and Ecology*. Their recent book *Ecology and Religion* (2014) brings many of the insights found in the longer series to the general public in a highly accessible form.[10]

In her article "The Emerging Alliance of Religions and Ecology," Mary Evelyn Tucker details approaches in which the major religions and indigenous traditions of the world can stimulate ecological and social change in liberating ways. Established religions have the institutional capacity to promote ethics and responsibilities by grounding people in nature's rhythms and ecological relations. Religious leaders are highlighting environmental problems in countries, seas, rivers, forests, and deserts throughout the world. Church members are being educated in the complexities of environmental degradation and the need to restore ecological communities. Worship ceremonies within Sunday services can awaken people to the needs and methods to preserve God's earth and personal salvation, and social activities beyond church walls can promote climate justice, peace, and ecological integrity. Organizations that promote conferences and interfaith services include the Center for Earth Ethics at the Union Seminary in New York City, GreenFaith in New Jersey, Earth Ministry in Seattle, and Faith in Place in Chicago.[11]

In April 2007, the University of Florida at Gainesville hosted the inaugural conference of the International Society for the Study of Religion, Nature, and Culture with the goal of promoting "critical inquiry into the relationships among human beings and their diverse environments, cultures, and religions." The conference was attended by scholars from around the world who discussed ways in which new religious and spiritual engagements with nature and culture can help to resolve environmental problems such as that of climate change.[12] The society's journal, *Religion, Nature, and Culture,* began publication in March 2007 in order to investigate questions such as: What are the relationships among human beings and what meanings are given to the terms *religion, nature,* and *culture*? What constitutes an ethically appropriate relationship between our own species and the places, including the biosphere, that we inhabit?

Among Catholics, Pope Benedict XVI urged bishops, scientists, and politicians to "respect Creation" while "focusing on the needs of sustainable development." To achieve this, Benedict recommended putting climate change on the top of the agenda. On April 26–27, 2007, the Vatican hosted a conference on climate change and development. Organized by the Pontifical Commission on Justice and Peace, it involved some forty participants and forty observers. It included scholars, scientists, and environmental ministers as well as bishops of the Catholic and Anglican Churches and representatives of Catholic religious orders and other ecclesial bodies representing some twenty countries.[13] Moreover, a number of theologians and clergy have spo-

*Figure 4.2. Pope Benedict XVI (b. 1927)*

ken about the need for an encyclical as well as an ecumenical statement on the environment from the Christian churches.

In May 2017, Benedict's successor, Pope Francis, presented President Donald J. Trump, on his visit to the Vatican, with a signed copy of his 2015 *"Laudato Si,"* "On Care for Our Common Home," which called for science and religion to join together to combat climate change.[14] Climate change action is part of the Pope's long-range effort to save rain forests in Latin America and encourage sustainable development throughout the world. Pope Francis takes his name from St. Francis of Assisi, the patron saint of animals and ecology who lived in caves, on mountainside, and in hermitages, praying for all creatures of the world. In June 2017, however, President Trump announced that the United States would withdraw from the Paris Climate Accord that had been reached in 2015. The United States cannot officially withdraw until 2020, however.[15]

Rosemary Radford Ruether, a radical Catholic theologian, challenges corporate globalization within developing countries in favor of local, small-scale enterprises that are environmentally sustainable. Large landowners and wealthy companies exploit the resources of the poor and extract their labor in an effort to claim profits. Changing the ways in which development works in the Global South is the first step toward an ethical, religiously responsible means of improving the lives of the oppressed. Green and renewable sources of energy, redistribution of food, the use of local resources for subsistence, and reliance on spiritual resources are methods by which developing peoples can improve their standards of life. Combating climate change must be integrated with spiritual commitments to the improvement of life for disadvantaged peoples.[16]

Indigenous American leader Winona LaDuke, Anishinaabe activist from the White Earth Reservation in northern Minnesota and executive director of Honor the Earth, has been a leader in combating fossil fuel development, especially on Native American reservations. Among many other Native Americans, she was outspoken about the (subsequently approved) extension of the Keystone pipeline from Canada to Texas across the Standing Rock Sioux Reservation in North Dakota. She stated that "it's time to move on from fossil fuels.... Each day or each week, there's some new leak, there's some new catastrophe in the fossil fuel industry, as well as the ongoing and growing catastrophe of climate change." Moreover, she continued, "from my perspective ... these guys don't need a pipeline. What they need is solar. What they need is wind.... they have ... class 7 wind out here. What they need is solar on all their houses, solar thermal. They need housing that works for people. They need energy justice."[17]

Indigenous peoples around the world consider mountaintops (which are often covered with glaciers) sacred places. Seeing the snows melt away and the glaciers disappear as a result of climate change is traumatic for many cultures. Glaciers can be either destructive forces or life-giving powers that need to be propitiated with ceremonies. Mountain deities responsible for life-giving waters may be abandoning their peoples. Villagers must now take greater care to preserve the ice that provides their source of water by restricting access and preventing the harvesting of great chunks of ice. New practices consistent with ancient lifeways and rituals are needed in order to maintain the very existence of mountaintop cultures and communities.[18]

## Eastern Religions

In many ways the goals of the religious groups discussed above resonate with the much older metaphysical beliefs of ancient Asia: Taoism, Buddhism, Zen Buddhism, Hinduism, Confucianism, and the many sects and traditions within Chinese, Japanese, and Indian thought. Can the beliefs of Eastern religions offer personal actions and ethics as guidelines for environmental ethics that would move the world out of the crisis of climate change and toward sustainability?

Eastern religions offer ways of thinking rooted in energy as a fundamental precept and therefore hold possibilities for a world based on process, change, and renewable "green" energy. Individual actions toward other humans can be expanded to embrace collective actions for the benefit of the planet. Taoism (or Daoism) goes back to the fourth century BCE and is based on "The Way," characterized by spontaneity, change, and compassion. Buddhism arose in India during the sixth to fourth centuries BCE; the Buddha achieved enlightenment through meditation and spent a life dedicated to alleviating human suffering. Zen Buddhism emphasizes meditation, self control, and introspection to achieve awareness and mindfulness. Hinduism originated in India around 500 BCE to 300 CE. The dharma is the right way or the eternal way, achieved through moral duties and fulfillment of obligations. Confucianism emerged from the thought of the Chinese philosopher Confucius (551–479 BCE). Humans are fundamentally good and can be taught to live properly, cultivate virtue, and to perform good deeds.

The main questions for the ecological humanities are: Have these religions in the past acted, and can they in the future act, to support nonexploitative ways of living and the restoration of the environment for future generations? But despite their emphasis on moral actions to benefit self and others, many scholars believe that in past centuries, the actions of Asian governments have not been environmentally positive. China, according to Joachim Spangenberg and Mark Elvin, is one of the worst polluters of the environment. Japan is an importer of fossil fuels and a major polluter of the oceans. India's air pollution, water pollution, and garbage accumulations are among the highest in the world.[19] Nevertheless, the traditional ethics of Asian countries can offer environmental ethics and behaviors that will help the world reverse the problems of climate change and attain sustainability in the twenty-first century.

Eastern philosophies are rooted in underlying concepts of change, process, energy, and transformation similar to those of some Western philosophies. The philosophies of Taoism and Confucianism, in particular, can be used in environmentally ethical ways to move us out of the Age of the Anthropocene and toward the Age of Sustainability. Taoism offers an alternative approach to knowledge, ethics, and the study of nature and the environment. As such, it is not only process-oriented but ethically relevant to the climate crisis. The Tao is the world's underlying energy.[20] In the sixth century BCE in China, the "Old Master," Lao Tzu, set down a collection of classic aphorisms known as the *Tao Te Ching* or *The Way* (the more modern Pinyin rendering is Daodejing or Dao De Jing). Lao Tzu was a contemporary of Confucius (who developed a philosophy of practical ethics). Over the centuries Lao Tzu and his Taoist philosophy became associated with "the people," while Confucianism appealed more to China's bureaucratic elite. By the end of the sixth century CE, Taoism was established as a popular cult, infused with alchemy, healing, popular magic, and ultimately with scientific developments such as the magnetic compass and gunpowder.[21]

The Tao, or The Way, is the ultimate reality, the One that underlies the appearances. As a cosmic process, it is the way of the universe. Taoists emphasize changes and flows within the whole, observing patterns within the cyclic, ceaseless motion of going and returning, expansion and contraction. Human intellect can never fully grasp the Tao, but people can observe nature to discover its ways. Its nonanalytic, intuitive, scientific approach achieves insights into transformation and change, growth and decay, and life and death through observation of the natural world. Taoist method links opposites, stressing contrary aspects, innate tensions, and spontaneity. Thus yin and yang are polar opposites within constant change. Yang represents the active, yin the receptive; yang is sunny, yin is shady; yang is light, yin is dark; yang is male, yin is female; yang is firm, yin is yielding; yang is heaven, yin is earth, and so on. The body is a balance between yin and yang, inside and outside, back and front. The Ch'i (qi) is its vital energy, the continuous flow that connects yang organs by way of yin meridians.

As in the postclassical process sciences, the Tao is the world's underlying energy. "What the Tao produces and its energy nourishes, nature forms and natural forces establish. On this account there is nothing that does not honor the Tao and reverence its energy." The Tao "produces, but keeps nothing for itself; acts, but does not depend on its action; increases, but does not insist on having its own way. This indeed is the mystery of energy."[22]

Ideas of energy and change as a give-and-take are fundamental to dealing with the implications of the Anthropocene. Flow and motion are the bases of the new forms of renewable energy, such as wind and water, and the absorption of the rays of sunlight in the solar panels that can in the future replace fossil fuels. A process-based world in which individuals, groups, and communities come together in dialogue to effect sustainable change is a better model than digging into the earth for coal and oil.

Confucianism, as a practical philosophy, is a basis for an ecological way of life that places the individual within the larger society and cosmos. One must cultivate one's own moral virtues so as to bring oneself into relationship with the larger whole. Nature is a moral good, a unity comprising relations and processes in which humans, the earth, and the cosmos form a triad. Life renews itself over and over in an act of flourishing and decay. Humans must take care of nature so as not to produce imbalances. On a practical level, they must provide for housing and human needs, carefully planting, harvesting, and storing grain while irrigating the land and conserving water. Humans must make every effort to offset pollution of water, soil, and air. In an effort to balance growth and progress, green technologies and alternative energy systems are essential for the future of the earth. These ideas could become the basis for an environmental ethic of sustainability that deals with the problems of climate change, resource depletion, and pollution.[23]

In the West, the connections between energy and process philosophy, as articulated by British philosopher Alfred North Whitehead (1861–1947), and process theology, as developed by California theologians John Cobb and David Ray Griffin, offer new rays of solar spirituality. Process philosophy owes its origins to Whitehead, who taught at Harvard University, and to Charles Hartshorne, Cobb's professor at the University of Chicago. Just as energy is the basis of the universe, so "process is fundamental. It does not assert that everything is in process ... but to be *actual* is to be a process." Process philosophy challenges the mechanistic idea that an atom or molecule remains fundamentally the same regardless of its relations. Instead, atoms acquire diverse properties in diverse relationships (or contexts). Atoms acquire different properties in different molecular arrangements because the new structures are new environments. Process philosophy thus substitutes an "ecological" (energy-based) theory of internal relations, in which entities are qualitatively changed in interactions, for the billiard ball model, in which entities are like machines—independent and unchanged, affecting each other only through external relations. Atoms and molecules therefore should be viewed not as machines that can be manipulated and controlled, but as ecosystems within which one must live and interact.[24]

Process theology thus is consistent with an ecological attitude and hence with a movement away from the domination of nature inherent in the Anthropocene in two ways: (1) its proponents recognize the "interconnections among things, specifically between organisms and their total environments"; and (2) it implies "respect or even reverence for, and perhaps a feeling of kinship with, the other creatures." Cobb and Griffin argue that process philosophy implies an ecological ethic and a policy of social justice and ecological sustainability: "The whole of nature participates in us and we in it. We are diminished not only by the misery of the Indian peasant but also by the slaughter of whales and porpoises, and ... the 'harvesting' of the giant redwoods. We are diminished still more when the imposition of temperate-zone technology onto tropical agriculture turns grasslands into deserts that will support neither human nor animal life."[25]

In his 2015 book, *Unprecedented: Can Civilization Survive the $CO_2$ Crisis?* David Ray Griffin applied his ideas on process theology to the climate crisis. He argues that it is essential for humankind to move away from fossil fuels and that through the development of clean energy systems, we can reach 80 percent clean energy by 2035 and 100 percent by 2050. To reach this goal, it is critical for the United States to lead the way.[26]

For Arkansas theologian Jay McDaniel, a protégé of Cobb and Griffin, the entire physical world has intrinsic value. Atoms, as individual things, have intrinsic value. Rocks express the energy inherent within their atoms. They too have intensity and intrinsic value, albeit less than that of living organisms. Outer form is an expression of inner energy. The assumption that rocks have intrinsic value, however, does not mean that rocks and sentient beings would necessarily have equal ethical value, but rather that they would all be treated with reverence. This could result in a new attitude by Christians toward the natural world, one that involves both objectivity and empathy and hence will be a powerful means of overcoming the worst effects of Anthropocene destruction.[27]

In his article "Process Philosophy and Global Climate Change," McDaniel writes: "Wherever we live, we can help build post-petroleum 'transition' communities that are compassionate, equitable, participatory, resilient, ecologically wise, and spiritually satisfying, *with no one left behind*. It can be a lot of fun."[28] Furthermore, McDaniel urges: "We can campaign *against* the fossil-fuel industry (350.org) and corporate powers which elevate greed to a virtue and denigrate compassion, and *for* a global power shift in which local people are empowered to make decisions that affect their lives. If we belong to an Abrahamic religious tradition (Judaism, Christianity, Islam) we can understand this engagement as an embodiment of the prophetic impulse, which critiques the principalities and powers, and works toward an alternative social order."[29]

## Religion and Actions to Combat Climate Change

What are some further examples of how spiritual beliefs can inspire actions to counteract climate change? Are there instances that offer hope for sustainability? In November 2017, in conjunction with international climate talks in Bonn, Germany, many groups traveled throughout Europe raising consciousness about the connections between climate change and sustainable development. Among them were indigenous leaders from Latin America, for whom "the land is sacred." According to Candida Dereck Jackson, vice president of the National Indigenous Alliance in Honduras, "We have been looking after the forests for thousands of years. We know how to protect them." These groups call for respect for land rights, recognition of crimes against the environment, direct negotiations over forest protection, decriminalization of indigenous activists, and the requirement that informed consent be obtained before any development by outsiders.[30]

The U.S.-based Interfaith Power and Light Campaign considers itself a "religious response to global warming." Comprising some twenty thousand congregations in forty states, it assists churches and other religious communities to cut down their carbon emissions, install solar panels, and educate congregants. The groups also have undertaken a major effort to divest from fossil fuel–based corporations. According to Joan Brown, executive director of New Mexico's Interfaith Power and Light, "Climate change is the biggest ethical, moral and spiritual challenge of our day."[31]

As the Reverend Brooks Berndt of the United Church of Christ puts it, "What actually motivates people is what I have found are the three great loves. Love of neighbor: You're aware of the real and present suffering climate or pollution are causing right now, so you're motivated once you have that awareness. Another is love of creation, concern for how our natural world is being decimated, extinction of animals, the acidification of the ocean and deforestation. The number one motive I've come across—although it's not for everyone—is love of children."[32] The Reverend Berndt argues that faith communities have a rich language to address issues of ethics and justice and that can draw people into seeing the injustices that are being wrought by climate change on fellow human beings and other living creatures.

Muslim communities are likewise taking action to combat climate change. Ahmed Bouzid, head of energy efficiency for Morocco, says that the Koutoubia Mosque, one of the oldest buildings in Marrakech, is now fully powered by solar energy. Installed in November before the United Nations climate summit in January 2017, an array of solar panels extends along the entire roof to the edge of the minaret. Six hundred additional mosques are scheduled to follow that example in the coming years. In this way, when people attend the mosques, they will open their hearts to change.[33]

The environmental network Caring for Creation makes the argument that the earth is God's gift to humanity and therefore humans need to take care of it. To take care of the earth is to support its life. Caring for creation nurtures humanity's relationship with God. As Dr. Katherine Hayhoe puts it, "Climate change disproportionately affects the poor and vulnerable—the very people that Christians are called to care and love."[34]

The wisdom of the world's great religions and the teachings of indigenous peoples were incorporated into the Earth Charter of the United Nations, a final version of which was released in 2000 for endorsement by the U.N. General Assembly. A follow-up to the Rio de Janeiro Earth Summit of 1992, the Earth Charter sets out goals and principles for a sustainable future. It responds to the fact that humanity is now at "a critical moment in Earth's history" and must choose to "form a global partnership to care for Earth and one another or risk the destruction of ourselves and the diversity of life."

The preamble to the Earth Charter includes a section titled "Earth, Our Home": "Humanity is part of a vast evolving universe. Earth, our home, is alive with a unique community of life. The forces of nature make existence a demanding and uncertain adventure, but Earth has provided the conditions essential to life's evolution. The resilience of the community of life and the well-being of humanity depend upon preserving a healthy biosphere with all its ecological systems, a rich variety of plants and animals, fertile soils, pure waters, and clean air. The global environment with its finite resources is a common concern of all peoples. The protection of Earth's vitality, diversity, and beauty is a sacred trust."[35] These goals, as endorsed by the world's peoples, must continue to inspire action and confidence that the earth can be preserved and sustained as a healthy place for all of life.

Spirituality can be a major means of dealing with climate change in the Age of the Anthropocene. From mainstream Judeo-Christian religions to, Islam, Far Eastern forms of spirituality, and the beliefs and practices of indigenous people around the world, changes in energy sources can be made. Ideals such as the good of the earth and of all people, especially the poor and impoverished, are the ethical bases for action. If more and more people bring their spiritual beliefs into play to make changes in their own communities, there is hope that by the mid-twenty-first century the world may have turned the corner toward renewable energy and a new Age of Sustainability.

# 5
# Philosophy

Twenty-first-century humanity is in the throes of a paradigm shift, one that is triggered by the rise of the new sciences of chaos and complexity, with climate change the most widespread catastrophe for the human future. Here I look at the prehistory of chaos and complexity in the Western world and the problem of nature as an unruly, unlawful, and unpredictable force, exemplified by earthquakes, volcanoes, tsunamis, and plagues. I focus on climate change in the Anthropocene as both global in scope and cumulative in effect and the ways it reflects uncertainties and limits to predictability. I show that new ways of dealing with the Anthropocene are critical for the future.

### Plato at the Googleplex

On December 19, 2017, Plato and I visited Google Headquarters in Mountain View, California. At 11:30 a.m., I arrived at Google's parking lot, where automobile parking was extremely limited owing to several rows of electric car charging stations, bicycle racks, and signs stating, "Reserved for expectant mothers" (with pictures of storks carrying babies in slings).

Under one arm, I was carrying my copy of Rebecca Newberger Goldstein's *Plato at the Googleplex* with which to engage my hosts John Markoff and Hans Peter Brondmo; over the other arm was a bag of articles on philosophy.[1] Plato's perspective was, after all, critical to

*Figures 5.1 and 5.2. Google Headquarters parking lot*

understanding the relationships between *information,* as understood in the current age of data analysis, and *knowledge,* as understood in fifth-century Greece. Both ways of knowing are essential to the age of the Anthropocene.

In Rebecca Goldstein's delightful book on "why philosophy won't go away," Plato (427–347 BCE) visits Google Headquarters with a tour guide for the purpose of promoting his latest book to Google workers. While waiting to give his talk, Plato (who is wearing a toga) and his tour guide Cheryl have coffee with a Google engineer named Marcus. In the course of the conversation, Cheryl explains to Plato that if you want to know anything you can simply "Google" the topic. Plato is incredulous that all knowledge can actually be located here at Google Headquarters. No, Cheryl explains, the knowledge isn't actually here at Google, but rather is stored "in the cloud." At this point Plato becomes very excited and wants to know more.[2]

The cloud symbolizes Plato's real world of pure forms or ideas, exemplified most powerfully for Plato by mathematics and subsequently by today's cloud computing. The cloud is opposed to the cave on whose walls only the shadows of real objects and hence the appearances of the pure forms can be seen.[3] In ancient Greece, the idea of the cloud as a symbol for the world of pure forms is associated with Plato's teacher Socrates. In fact, the cloud as associated with knowledge and the process of reasoning is ridiculed by the Greek playwright Aristophanes (b. 750 BCE), who depicts Socrates descending from the clouds in a basket.[4]

Not only is Plato excited that knowledge can be stored in the cloud, he also wants to know how it is possible to search for all that knowledge. Marcus, the engineer, elaborates that, in response to questions posed by users, the search engine finds thousands of results and then sorts them into a manageable order. Google is gathering

*Figures 5.3 and 5.4. Carolyn Merchant's visit to Google Headquarters, carrying* Plato at the Googleplex, *December 19, 2017*

PHILOSOPHY

*Figure 5.5. Socrates descending from the clouds in a basket, sixteenth-century engraving*

knowledge, Marcus states smugly, and—as Plato himself knows—knowledge is a good thing. At this point, however, Plato lowers his head and very softly whispers, "It's information, not knowledge."[5]

Toward the end of Plato's visit, the question is raised as to whether Google can supply answers to ethical questions. Can ethical issues be resolved by a search engine that assigns different weights to people's votes on particular moral dilemmas? If most people think a life of indulgence is what constitutes the good life, then the search engine will come up with that particular answer. Thus questions such as "What is a life worth living?" would be answered not by ethical experts (as in Plato's *Republic*), but by an "Ethical Answers Search Engine"—EASE. Mathematics, according to Marcus, is the method used to set up ethical answers by crowd-sourcing. But as they are walking (which, of course, is how Plato does his best thinking—peripatetically) toward the lecture hall, Plato subtly guides Cheryl to the recognition that ethical questions "cannot be answered with ease." Nevertheless, the final question remains: "Is that with EASE, spelled with all capitals, or just lowercase 'ease'?"[6]

The Googlers who come to hear Plato's book talk are wildly enthusiastic. In fact, they are all wearing T-shirts "showing two guys in togas, one with his finger pointing up and the other with his hand out flat." The first, of course, is Plato, whose finger points upward to the pure forms as the sources of abstract knowledge. The other is Aristotle, for whom forms are imbedded in the material objects themselves.[7]

The T-shirt in fact depicts the fundamental dilemma of the Anthropocene. If we are living in a Platonic world of mathematics described by ones and zeros and controlled by computers and search engines in which knowledge is stored in the "cloud," then do humans, as anthropos, have the ultimate power to control nature? If we can predict the future by the use of mathematical equations and computer algorithms, can we then control the fate of nature and hence of humanity? Can we eventually make the ethical choices and decisions that will curtail the use of fossil fuels and halt the concentrations of greenhouse gases in the atmosphere?

If, on the other hand, knowledge (as in Aristotle's forms) is embedded in the material world and that world is changing as a result of human inputs from carbon dioxide and greenhouse gases, then humanity may be paving the way toward its own ultimate death—a new "death of nature" that now includes humans. Is there an ethical solution to this dilemma? Can humanity make philosophical and political choices that will lead to a sustainable future and not an anthropogenic demise? Early Greek and Roman philosophers give us clues as to how to rethink philosophy for a new Age of Sustainability in the future.

*Figure 5.6. Plato and Aristotle, in Raphael's* The School of Athens, *in the Vatican*

## PHILOSOPHY

### Rethinking Philosophy in the Age of the Anthropocene

The philosophers of ancient Greece and Rome raised fundamental questions about the human relationship to the external world:

1. The ontological question: What is the world made of and how does change occur?
2. The epistemological question: How do we know?
3. The ethical question: What ought we to do?

Although people of all periods and ages have asked and answered these questions, they are particularly relevant for the current Anthropocene era. If the Anthropocene is the result of the human control of nature through science and technology, such that greenhouse gases from the burning of fossil fuels are discharged into the atmosphere, how can we rethink the history of philosophy through the eyes of the anthropos?

Among those philosophers most relevant to the anthropocenic future are the sixth- and fifth-century BCE "naturalists" of Asia Minor. Unlike the philosophers of ancient Mesopotamia, who attributed change to the actions of gods such as Marduk (god of thunder, water, vegetation, and magic), the Milesians answered the question "What is the world made of?" by naming material elements as first principles and identifying universal laws to predict events. For Thales of Miletus, an Ionian philosopher, who flourished circa 585 BCE in Asia Minor, the answer was "All is water." Rain falls to the earth, where it helps plants to grow and sustain life, and then evaporates into clouds that once again produce the rains that water the earth. The earth, Thales said, was a circular disc floating on water. Other Milesians, such as Anaximenes (fl. 546 BCE), argued that the first

*Figures 5.7 and 5.8. Heraclitus of Ephesus (540–475 BCE) and Parmenides of Elea (fl. 504 BCE)*

principle was air, while Anaximander, also from Milesia, held that it was the Infinite.[8]

In addition to the material components that provide answers to the ontological question "What is the world made of?" the question of change or process is equally fundamental to the future of human life in the Anthropocene. In the Greco-Roman world, that question was first addressed by Heraclitus of Ephesus. His major contribution was the statement "All is change." Everything is in flux. "You could not step twice into the same river; for other and yet other waters are ever flowing on." The only constant is the fact of change itself. This is the concept of the dialectic later articulated by Hegel as idealist dialectics and by Marx and Engels as materialist dialectics. The idea of change as a dialectic—a back and forth between humans and nature—is critical to whether the airs and waters of the twentieth century have been fundamentally altered by humans in potentially irreversible ways through the influx of greenhouse gases.[9]

Another Greek philosopher critical to the future and hence to the Anthropocene is Parmenides of Elea in southern Italy. His contribution (in contradiction to that of Heraclitus) was to assert that the world is made of being and that there is no change at all. If this were true, then nothing could evolve and there would be no hope for a better world. Parmenides framed his argument as "Being is and Not-Being is not." This statement is the law of identity: a is a, which is the basis of logical reasoning and of mathematics itself. He likewise framed the law of noncontradiction: you cannot say "a is not-a." Logic is identical with thinking, whereas opinion equals nonthinking. Parmenides's logic is the basis of mathematics, hence of the digital world of ones and zeros in which we live today and which is the very foundation of the Googleplex.[10]

The "atomists," such as Empedocles of Akragas in Sicily, Anaxagoras of Asia Minor (510–428 BCE), and Democritus of Abdera in Thrace, thought in terms of material particles that moved through void space, boldly contradicting Parmenides' argument that "Being is and Not-Being is not." Not-being exists and is called the void or space. Particles, or atoms, move through void space, thereby allowing change to occur. For Empedocles the processes of love and strife acted on the four elements (earth, air, fire, water) to create change.[11] Democritus introduced the concept of quantitative atomism. The word *a-tom* means "not-cuttable." Not-being thus exists and it is called the void, or space. "To be" means to be an atom or to be a space (a void). Atoms are in ceaseless motion. Change is the union and separation of atoms.[12]

Following these early Greek materialists, by the seventeenth and eighteenth centuries the atoms of the Greeks became the corpuscles of René Descartes, the atoms of Thomas Hobbes, the hard massy particles of Isaac Newton, and ultimately the atoms and molecules

## PHILOSOPHY

*Figures 5.9 and 5.10. Empedocles of Akragas (fl. 444 BCE) and Democritus of Abdera (b. 450 BCE)*

of the industrial era.[13] The Anthropocene depends on seeing the material world as comprising the carbon atoms that make up the carbohydrates of living organisms and the dead organisms that decay into the fossil fuels that are burned today.

Together, therefore, Heraclitus's concept of change and Parmenides's concept of logic are the foundations for both an understanding of and a way out of the Anthropocence. Change defines the physical world in which we live; logic and numbers define the real world of mathematics. Numbers can describe the physical world, although only imperfectly. The philosophers of the ancient world discovered both the beauty of mathematics and its limitations. As elaborated by Pythagoras of Samos, the cosmos is based on numbers, and numbers are eternal and unchanging. Numbers are being, substance, matter. Numbers are sacred and eternal. Numerical proportions produce

## PHILOSOPHY

Figure 5.11. Pythagoras of Samos (ca. 578–510 BCE)

music and the harmony of the spheres. Musical notes are numerical ratios that constitute 2:1 = octave; 3:2 = fifth; 4:3 = fourth. The planets in their circuits around the earth emit musical notes.[14]

The Pythagorean theorem, brought from Egypt (and/or Mesopotamia) by Pythagoras, is the beautiful relationship that states that the sum of the squares on the legs of a right triangle is equal to the square on the hypotenuse ($a^2 + b^2 = c^2$; for example, $3^2 + 4^2 = 5^2$). The numbers are all whole numbers and rational. But the Pythagoreans soon discovered a terrible unmentionable truth that upset the foundations of their entire cosmic system. The diagonal of a unit square (in which each side = 1) is irrational; that is, the square root of 2 = 1.414214 ... ad infinitum. The world is therefore both rational and irrational, logical and illogical.[15]

Parmenides and Plato provide the ontological foundations of the digital universe epitomized by the Googleplex and by the anthropocenic idea that humans can control nature through the predictability of logic and mathematics as developed by the Pythagoreans and others. Conversely, the early atomists and Aristotle supply the idea that the universe is material and controllable via technology. But what about the Heraclitean idea that the universe is fundamentally active, changing, and often unpredictable—the irrational element of the Pythagorean heresy? Is there a way to rethink the Anthropocene so that humans can interact with a complex changing nature? Is there a way of slowing greenhouse gas emissions and moving toward a renewable energy economy that goes beyond the Anthropocene?

### Mechanistic Science and Predictability

The Anthropocene as an era in which humans could control nature became a significant problem in the modern world. The Scientific Revolution of the seventeenth century brought together the astronomical theories of Nicolaus Copernicus, Tycho Brahe, and Johannes Kepler and the terrestrial mechanics of Galileo, Descartes, and Robert Boyle. Isaac Newton's *Principia Mathematica* of 1687 synthesized the laws of mechanics into a unified system that described the three laws of motion and the law of gravitation. Experimentation and technology coupled with mathematics allowed humanity to change the planet for the benefit of the anthropos.[16]

Newton's contemporary and competitor Gottfried Wilhelm Leibniz, who debated with Newton over the nature of God and co-invented the calculus, also invented a calculating machine that became the basis of the computer. Building on earlier inventions such as Blaise Pascal's adding machine, Leibniz's 1671 *instrumentum arithmeticum* (later called a "stepped reckoner") allowed for multiplication and division as well as addition and subtraction, thereby enhancing the idea of predictability through the use of mathematics. Leibniz

PHILOSOPHY

*Figures 5.12 and 5.13. Isaac Newton (1642–1727) (left), portrait by Godfrey Kneller, 1689, and Gottfried Wilhelm Leibniz (1646–1716), copper engraving, 1775, by Johann Friedrich Bause, after a 1703 painting by Andres Scheits*

also invented the digital binary code consisting of ones and zeros (as an alternative to the decimal system based on the numbers one through ten) that is the basis of today's computer coding.[17] Charles Babbage's "Difference Engine" of 1822 and his "Analytical Engine" of 1842 furthered the pathway toward the twentieth-century computer and the digital world that has transformed the lives of twenty-first-century citizens.

Today's computers operate on the basis of digital data—Leibniz's binary code of ones and zeros—that attempt to describe the physical (or analog) world in which we live. The difference between the two worlds is once again the difference between the predictability of pure mathematics—the numbers and pure forms of Plato and the Pythagoreans—versus the changing material world of Heraclitus and

*Figure 5.14. Albert Einstein (1879–1955)*

the Greek atomists. While mechanistic science and its mathematical predictability work well in most physical situations, giving us confidence that we can cross bridges and fly in airplanes, the problems of unpredictability loom in infrequent occurrences—earthquakes, volcanic eruptions, tsunamis, plagues. Here nature is autonomous and often unpredictable, defying mathematical precision and computer coding and hence human control over the environment.

### Challenges to Mechanistic Philosophy and Predictability

In the late nineteenth and early twentieth centuries, Max Plank (1858–1947), Neils Bohr (1885–1962), Albert Einstein (1879–1955), Werner Heisenberg (1901–76), and others began to challenge mechanistic science and with it the possibility of human control over nature. The photoelectric effect is the idea that light can be thought of as quanta or particles of energy called photons, as well as waves. Mat-

*Figure 5.15. Edward Lorenz (1917–2008)*

ter was understood as having a substructure made up of electrons, protons, and neutrons. In 1905, Einstein's special theory of relativity stated that the velocity of light was the top limiting velocity in the universe. In 1927, Heisenberg's uncertainty principle stated that both the position and momentum of a particle could not be known simultaneously. These ideas challenged mechanism, but at a level not apparent in everyday life.[18]

During the 1970s and 1980s, however, challenges to mechanistic science and hence to the anthropogenic control over nature in the everyday world began to appear. In December 1972, Edward Lorenz presented a paper to the American Association for the Advancement of Sciences titled "Predictability: Does the Flap of a Butterfly's Wings in Brazil set off a Tornado in Texas?" This phenomenon became known as the "butterfly effect," or sensitive dependence on initial conditions. It applied especially to weather patterns as being chaotic, and hence unpredictable. Irregularity is thus a fundamental

## PHILOSOPHY

*Figure 5.16. Ilya Prigogine (1917–2003)*

property of the atmosphere. Most environmental and biological systems are in fact nonlinear and chaotic and cannot be fully controlled by human beings.[19]

The idea of unpredictability was pushed further by Nobel Prize–winning physicist Ilya Prigogine, who wrote *From Being to Becoming: Time and Complexity in the Physical Sciences* (1980). Prigogine and his coauthor Isabelle Stengers then wrote a more accessible version of his work, *Order out of Chaos: Man's New Dialogue with Nature,* in 1984. In classical thermodynamics, systems are in equilibrium or near equilibrium. These include stable systems such as pendulum clocks, steam engines, refrigerators, and solar systems. Here, small changes lead to adjustments and adaptations, described by the mathematics of calculus and linear differential equations. But in cases of large inputs, nonlinear relationships take over. In these far-from-equilibrium systems, small inputs can produce new and unexpected effects. Most systems, such as biological, ecological, and social systems, are open systems and not closed mechanical systems. Here, small perturbations can cause a breakup and reorganization of matter and energy.

New enzymes or new cellular structures can appear at the biological level, and at the social level new societies can emerge in response to a disruptive challenge.[20]

In 1987, James Gleick, an editor at the *New York Times,* popularized chaos theory in his book *Chaos Theory: Making a New Science.* He interviewed numerous scientists who had published on chaos theory and asked them to explain it so he could understand it. Chaos theory argues that most biological and ecological systems cannot be described accurately by the linear differential equations that pertain to mechanistic science and instead are governed by nonlinear, chaotic relationships. Gleick discussed the ways in which natural entities such as trees, coastlines, and snowflakes can be understood as self-similar patterns within patterns, or fractals.[21]

In 1992, Mitchell Waldrop wrote a similar book about the science of complexity, *Complexity: The Emerging Science at the Edge of Order and Chaos.* Waldrop interviewed scientists and social scientists at the Santa Fe Institute in New Mexico where integrative research was being conducted on the origins of complex structures from bacteria to galaxies and from early to advanced societies and economies.[22]

Yet another account of the differences between mechanistic and nonmechanistic systems appeared in Daniel Botkin's *Discordant Harmonies: A New Ecology for the Twenty-first Century* in 1990. Botkin examined metaphors used to describe the earth over time, including nature as a divine order, earth as a fellow creature, and nature as the great machine. Most significant for the idea of nature as a self-active, unpredictable entity was Plotinus's view of nature as a discordant harmony composed of the simultaneous movements of many tones: sometimes harsh, sometimes pleasing.[23]

The idea of the Anthropocene as a system of nature that can be controlled by human beings through mathematics, experimentation, and technology is thus being increasingly challenged by concepts of nature as chaotic, complex, discordant, and unpredictable. Such ideas contest the human ability to predict what will happen to the geology and ecology of the planet as greenhouse gases continue to be emitted into the atmosphere by the burning of fossil fuels. Unforeseeable, unpredictable, and uncontrollable outcomes are increasingly part of the future of human-nature relations and interactions. As Dara Khosrowshahi, CEO of Uber, puts it: "We are a digital company that is organizing the physical world. The physical world is a lot messier than the digital world: messy, unpredictable, tough to organize—and also more fundamental."[24]

In this new world of the twenty-first century and beyond, coding confronts chaos; the binary world is challenged by the analog world; computers are compromised by autonomous nature. The world of ones and zeros, of yeses and nos, of Plato's pure forms and imperfect appearances, is more complex and chaotic than either Newton or Leibniz could have anticipated.

While mechanistic systems that rely on linear equations, data systems, coding, and predictability are still of fundamental importance to life in the everyday world, the unpredictability of an active nature that is continually responding to anthropocentric inputs must increasingly be taken into consideration. Nature and humanity as interactive, changing, and exchanging entities must be an ever-increasing part of a new complex human-natural world. Plato must come together with Plotinus, Heraclitus with Parmenides, computers and data bases with the ever-flowing, ever-changing outlines of coasts, rivers, shorelines, and glaciers. More research beyond what I have attempted in this book is required. The role of philosophies in other countries and how they can influence the world of today is critical. A new ethic for the earth of the twenty-first century and beyond is desperately needed. That ethic, as the next chapter will show, is an ethic of partnership between humanity and the earth. The new era must move beyond the Age of the Anthropocene and human control to an Age of Sustainability and partnership between humanity and nature.

# 6
# Ethics and Justice

Ethics and justice are essential ingredients for a transformative response to the Anthropocene. Here I reevaluate ethical frameworks and approaches, such as egocentric (liberal), homocentric (anthropocentric), ecocentric (ecological), and multicultural theories, through the lens of the Anthropocene. I propose a new approach, one that I have called partnership ethics, as a way to bring humanity and nature into an interactive relationship that recognizes the needs of both nature and humans as a basis for mutual survival in the twenty-first century. I likewise examine the impacts of climate change on marginalized peoples and argue that new theories of justice are needed. Direct involvement by low-income communities, indigenous peoples, women, and peoples of color in all aspects of decision-making, planning, and policy is essential to mitigating the negative impacts of the Anthropocene.

## Environmental Ethics and the Anthropocene

An ethical response to the Anthropocene and climate change requires both an understanding of the major forms of environmental ethics and what new ethics are needed for life in the future. Environmental ethics link theory with practice. Thus an ethic that responds to the problems of the Anthropocene should also offer ways to resolve those problems.[1]

Figure 6.1. John Locke
(1632–1704)

An egocentric ethic is grounded in the self. It is based on an individual "ought" focused on individual "good." It involves the claim that what is good for the individual will benefit society as a whole and is based on a philosophy that treats individuals (or private corporations) as separate—but equal—social atoms. The egocentric ethic is grounded in the philosophies of Thomas Hobbes (1588–1679) and John Locke, in which individuals maximize their own self-interest and accept a set of rules that govern each individual's ethical commitment to live in an orderly society. It is the guiding ethic of private entrepreneurs and corporations whose primary goal is the maximization of profit, especially industries that use fossil fuels for production purposes. The Anthropocene as an era in which greenhouse gases are increasing exponentially is rooted in a corporate-based egocentric ethic.[2]

A homocentric (or anthropocentric) ethic is grounded in society. A homocentric ethic underlies the social interest model of politics and the approach of environmental regulatory agencies that protect

## ETHICS AND JUSTICE

*Figures 6.2 and 6.3. Jeremy Bentham (1748–1832), engraved by J. Posselwhite from an original picture by J. Watts, and John Stuart Mill (1806–73)*

human health. The utilitarian ethics of Jeremy Bentham and John Stuart Mill, for example, advocate that a society ought to act in such a way as to ensure the greatest good for the greatest number of people. The social good should be maximized, social evil minimized. In the Age of the Anthropocene, a homocentric ethic underlies the efforts of federal and state governments to regulate the amount of greenhouse gases entering the atmosphere and to offer health and human services to those suffering from lung diseases, skin cancers, and other consequences of air pollution.[3]

An ecocentric ethic is grounded in the cosmos. The whole environment, including inanimate elements, rocks, and minerals, along with animate plants and animals, is assigned intrinsic value. The ecoscientific form of this ethic draws its "ought" from the science of ecology. Modern ecocentric ethics were first formulated by Aldo Leopold during the 1930s and 1940s and published as "The Land Ethic,"

*Figure 6.4. Aldo Leopold (1887–1949)*

the final chapter of his posthumous *A Sand County Almanac* (1949). Leopold wrote: "A thing is right when it tends to preserve the integrity, beauty, and stability of the biotic community. It is wrong when it tends otherwise." With respect to the Anthropocene, an ecocentric ethic means that humans should endeavor to do everything possible to avoid ecological changes brought about by global warming, such as species extinctions, habitat transformations, and northward and southward migrations of increasingly maladapted species.[4]

In recent years a number of philosophers have moved beyond ecocentric ethics to ethical formulations that include principles of environmental justice and cultural diversity in response to globalization. J. Baird Callicott proposes a multicultural ethic that builds on the complementarity between biological diversity and cultural diversity. The human species is one species of many cultures. All humans are part of a local, bioregional culture and an international global culture. Grounding an ethic in postclassical science transcends the conflicts that may occur between local and global geopolitics. A multi-

## ETHICS AND JUSTICE

cultural ethic is particularly applicable to cases of climate justice, as elaborated below.[5]

As I elaborate further in the epilogue, my own view is that we need not only a new ethic but also a new Age of Sustainability that will supersede the Age of the Anthropocene. My own ethic is an ethic of partnership between humans and nonhuman nature. It states: *A partnership ethic holds that the greatest good for the human and nonhuman communities is in their mutual living interdependence.*[6] A partnership ethic is grounded not in the self, society, or the cosmos but in the idea of relation. It has five precepts:

- Equity between the human and nonhuman communities.
- Moral consideration for both humans and other species.
- Respect for both cultural diversity and biodiversity.
- Inclusion of women, minorities, and nonhuman nature in the code of ethical accountability.
- An ecologically sound management that is consistent with the continued health of both the human and the nonhuman communities.[7]

A partnership ethic entails a viable relationship between a human community and a nonhuman community in a particular place, a place in which connections to the larger world are recognized through economic and ecological exchanges. It is an ethic in which humans act to fulfill both humanity's vital needs and nature's needs by restraining human hubris.

How can environmental ethics help to resolve problems of anthropogenically caused climate change through climate ethics and climate justice?

## Climate Ethics

Leading scholars, as well as politicians and scientists, have argued that ethics is not just important to the resolution of climate change, it is *the* principal factor needed to manage global warming. "Natural, technical and social sciences can provide essential information and evidence needed for decisions on what constitutes 'dangerous anthropogenic interference with the climate system.' [But] at the same time, such decisions are value judgments," notes the IPCC.[8] New principles are needed to help determine which parties are responsible for climate change mitigation and to what degree. New atmospheric targets are required to account for differences in previous and current emissions, wealth and poverty, quality of life, and stages of industrial development. Given that in poor countries, carbon is used mostly for necessary activities such as cooking and home heating while in the industrialized nations it is used predominantly for activities such as driving, flying, and heating water, differences in carbon needs must be evaluated.

Because of the gap between need and will, ethics is an essential ingredient for a productive response to climate change. Every major aspect of climate negotiation is an ethical issue; ethical principles and reasoning are needed in order to work through each of the challenging issues on the negotiating table: responsibility for damage, reasonable targets, allocation of carbon emissions trading, costs to national economies, degrees of responsibility, assessment of new technologies, and procedural fairness. Existing ethical theories need to be evaluated and new theories proposed for resolving the complex issues associated with climate ethics.

According to philosopher Peter Singer, we must "see the atmosphere as a resource for which we are all responsible; we must agree

*Figure 6.5. Peter Singer (b. 1946)*

on how responsible each party is for protecting the atmosphere and who must pay how much to protect it." Singer makes an analogy between two hundred villages overfishing a nearby lake and two hundred nations overpolluting the atmosphere on which we all depend. The best way to understand this ethical problem, he says, is to think about how best to divide a scarce resource that no one owns—in this case, the atmosphere or, more specifically, "the capacity of the atmosphere to absorb our waste gases without changing the planet's climate in harmful ways."[9] Singer argues that from an ethical standpoint the developed nations, which are better off, should bear the largest burden of the costs of combating climate change.

In a foundational article in 2006, "A Perfect Moral Storm," philosopher Stephen Gardiner of the University of Washington, Seattle, pointed out that climate change comprises deep ethical challenges and is far more complex than the single-substance, single-industry changes such as the CFCs (chlorofluorocarbons) that were expand-

# ETHICS AND JUSTICE

*Figure 6.6. Stephen M. Gardiner (b. 1967)*

ing the ozone hole in the 1980s.[10] CFCs required the action of very few parties, and industry leaders and government regulators saw that replacement technologies were actually cost-effective.

But in the case of climate change, says Gardiner, we face three stark challenges: a vast dispersion of causes and effects, great fragmentation of agency, and the inadequacy of corporations, government, and science to deal with the issues. These three challenges demand ethical solutions. Thus, climate change "is a complex problem raising issues across and between a large number of disciplines, including the physical and life sciences, political science, economics and psychology. But without wishing for a moment to marginalize the contributions of these disciplines, ethics does seem to play a fundamental role."[11]

Because almost everybody on earth uses and emits fossil fuel, we need to build much stronger institutional capacity to deal with alternative energy sources. In the absence of individual compliance, we need a strong system of global governance. If we see climate change as posing a problem, Gardiner says, then we see that our actions that force climate change are open to moral assessment. This leads to the need for "some account of moral responsibility, morally important interests, and what to do about both. And this puts us squarely in the domain of ethics."[12]

## Climate Justice

The term *environmental justice* (EJ), which is directly linked to and expands into the climate justice movement, came into widespread use in the early 1990s. It describes a movement born of a multitude of smaller movements created by minority and underprivileged groups in the United States and abroad. These groups address the unjust distribution of environmental burdens and benefits—such as the location of industrial facilities and their pollution and access to wealth, good food, clean air and water, and parks and recreation. Environmental justice is environmental ethics "on the ground." In turn, ethics playing out in the offices of social and political institutions is spurred on by climate justice issues at ground zero.

A number of examples nationwide that affected minority communities led to the environmental justice movement and to the related issue of climate justice.

- In 1982, African Americans and Native Americans staged a major protest over a landfill used for dumping polychlorinated biphenyls (PCBs) in Warren County, North Carolina. PCBs have been widely used throughout the country as insulators and are contained in many consumer products. The protest, documented in Eileen McGurty's *Transforming Environmentalism: Warren County, PCBs and*

*Figure 6.7. Warren County protest, 1982*

*the Origins of Environmental Justice* (2007), represented the beginnings of the environmental justice movement.

- In 1983, the City of Los Angeles proposed the Los Angeles City Energy Recovery Project, or LANCER, calling for a network of three garbage incinerators capable of dealing with sixteen hundred tons of garbage per day. Latin American activists Maria Roybal and Aurora Castillo, who founded MELA (Mothers of East Los Angeles) in 1984 to support Latina mothers who live in poor areas, staged protests against LANCER because of negative heath implications for minorities. After hearings, in 1990, the city abandoned the project and agreed to sell the site to a nonprofit group for a housing development project.[13]
- In 1989, an eighty-mile stretch of land between Baton Rouge and New Orleans became known as "Cancer Alley" when Dow Chemical Company bought out vinyl-chloride-contaminated neighborhoods occupied primarily by African American workers.

## ETHICS AND JUSTICE

*Figure 6.8. Protest over proposed sewage plant in West Harlem, 1990*

- In 1990 in West Harlem, New York, black women organized a protest over the North River Sewage Treatment Plant that was producing toxic emissions in their neighborhood. They founded WE ACT and took on additional environmental justice projects, such as clean air, transit, toxic-free products, sustainable land use, waste reduction, and open space.[14]
- In 2014, in Flint, Michigan, lead began leaking from water pipes into homes after the City of Flint changed its water supply from Lake Huron to the Flint River to save money. According to the Michigan Civil Rights Commission, Flint's African American majority suffered the most from the resulting lead poisoning.[15]

Minority groups and authors responded to these and numerous other examples with environmental analyses that led to the environmental justice movement and then the broader climate justice move-

Figure 6.9. Robert Bullard (b. 1946)

ment. In 1987, the United Church of Christ Commission for Racial Justice published a national report, "Toxic Wastes and Race in the United States," mapping the number of uncontrolled toxic waste sites in areas of 50 percent or greater and 20–49 percent Hispanic population.[16]

In 1990, African American activist Robert Bullard of Clark Atlanta University in Georgia wrote *Dumping in Dixie: Race, Class, and Environmental Quality* followed by *Confronting Environmental Racism: Voices from the Grassroots* in 1993. In 1994, he founded the Environmental Justice Resource Center and compiled a directory of over four hundred "people-of-color environmental groups." Most of these groups originated as social justice groups and then expanded to include environmental and climate justice issues. Bullard later moved to Texas Southern University in Houston, where he wrote *Black Metropolis* (2007). He identified the problem of "free land, free labor, free men:" the "free land" was actually taken from Native Americans; the "free labor" was performed by African American slaves; and the "free men" were men of property who had the vote.[17]

### Environmental Justice and Climate Justice

Climate change as an ethical issue greatly alters and expands the field of environmental justice. Climate change scholars call into question the tensions between homocentric, utilitarian approaches and egocentric, rights-based paths to equity, with the recognition that both concepts are necessary and must be brought into a common functional framework. Looking at how these basic ethical theories play out on the ground, scholars have noted that "the distinction between utilitarian and rights-based approaches to equity ... actually lies at the heart of the crisis of governance that pervades the local, national, and global communities.... Individual, local or ethnic rights ... ought not to be violated even at the expense of the aggregate good."[18] New theories of climate justice are required to meet emerging challenges, and new frameworks are needed for practice by groups on the ground that are immediately impacted by climate change.

Environmental justice reflects the direct involvement by marginalized peoples in the naming and claiming of how environmental and climate change issues impact their communities. The 2003 report of the U.S. Commission on Civil Rights states clearly that the direct involvement by low-income groups and communities of color in all aspects of decision-making, planning, and evaluation of environmental projects and policies is essential to any ongoing effort to mitigate the negative impacts of environmental hazards.[19] The environmental justice literature emphasizes the role that communities of color play in deciding what "justice" looks like. These groups can help to determine the character of climate justice and how it is applied through their focus on race, gender, and class differences that inform individual and community responses to outcomes. Thus new theories of climate justice can be defined by those most affected.

Environmental justice movements are likewise informed by academic ethics. Ethical frameworks must account for distributive justice, unequal access to resources, and unequal abilities to pay. Theories of justice should capture the complexities of climate inequities. Existing justice systems must respond adequately to climate change and new theories and practices must be developed.

Climate change is a special environmental justice issue in its global character. It synthesizes many smaller environmental justice movements, as so many of them are tied to questions of energy and equity. Almost across the board, the wealthiest nations have benefited from industrialization and may continue to live prosperously for some time, while developing nations face more immediate crises. Within the United States, however, equity issues are particularly poignant, as the U.S. has the largest gap between poor and wealthy classes of any industrialized nation. While the wealthy can make their personal wagers—insurance, mortgages, housing locations—against catastrophic climate change in the coming few decades, the poor face subsistence crises, diminishing resources, and flooding that can lead to immediate and catastrophic losses.

## Climate Justice and Native Peoples

Climate justice has different meanings for Alaskan natives, Native Americans, Latinos, and African Americans. These groups, facing harder burdens than white Americans and with fewer economic resources, have important history and knowledge that can contribute to mitigating difficult climate impacts. Across great differences in cultural perspectives, worldviews, and language, scholars and activists have begun identifying the ingredients necessary for climate adaptation and equity, including viable networks of social, cultural, political, and economic support. Native peoples from around the mainland United States, Hawaii, and Alaska have identified six key sectors critical to native peoples in the United States: "Water, agriculture, human health, wildlife and ecosystem loss, sovereign borders and boundaries, and tourism and recreation."[20] Active participation of Native American leaders in the climate policy process engages people highly affected by climate change, while bringing the strengths and cultural history of native groups to the policy table.

Two of the most heavily impacted groups within the United States, Arctic Native peoples and Native Americans of the lower forty-eight states, not only have immediate experiences of climate change but bring long-held practices, cultural views, and tools to understanding the issues involved. In the past few years Inuit villagers have seen large pieces of their shoreline break off and float away, and they are now moving hundreds of miles inland at a cost of tens of millions of dollars. But their actual story is much worse. In the immediate future, these native tribes face famine and bankruptcy, as energy prices rise and Arctic species that are staple foods lose habitat and face extinction. Nor is it only a question of certain key resources—we are losing the integrity of entire Arctic ecosystems as life-support systems.[21] Native groups aim to focus Congress's attention on "the unique needs of Native American communities,... the need for money for relocation efforts, the need for legislation that caps emissions, and the need for some kind of governmental entity that is responsive to these needs," according to Heather Kendall-Miller of the Native American Rights Fund.[22] Alaskan Native and American Indian concerns about global warming make visible the energy and subsistence crises faced by native peoples.

Indigenous peoples' responses to climate change reveal several dimensions of climate justice. Thus a tribal body might advocate a response to climate change that addresses a very local impact (for example, pressure on treaty-negotiated water reserves by non-native communities) in a way that is a regulatory intervention posed within a specific federal-tribal legal and ethical framework. But that response might also reflect the need for particular types of water for certain cultural and spiritual uses. Not all water is the same. Some is spiritually acceptable and some is spiritually contaminated. Hence a tribal response might be about ethics, justice, and religion simultaneously.

The list of climate-related concerns for poor African American and Latino communities is daunting: a much higher percentage of poor versus affluent groups live near toxic sites; a high percentage of them are in regions vulnerable to climate change catastrophes, such as urban centers and coastal regions. Moreover, in the coming decades, without major economic transformation, poor classes will pay escalating energy and food bills—and, compared to rich classes, a much higher percentage of their pay is already devoted to food and energy. Blacks experience much higher rates of cancer and asthma due to toxic pollution, and twice as many blacks as whites lack health insurance. Increasing impacts of climate change in the United States will quickly worsen existing gaps in wealth. The neglect and abuse of African Americans during and after the Katrina crisis poignantly illustrate the dimensions of justice that will continue to be at play under strengthening hurricanes associated with global warming. In responding to these issues, minority communities, those most immediately impacted by climate change, are making significant contributions to climate justice.

## ETHICS AND JUSTICE

Policy makers have begun to formulate concepts and frameworks for climate justice. Until recently, debate centered primarily on a limited schema, focused on allocating to each country equal per capita emissions; rights according to historical responsibility; rights according to a country's ability and willingness to pay; or some combination thereof. But climate scholars are increasingly including other justice issues. One major area is procedural justice, such as the role of developing countries in decisions on adaptation to climate change.[23] Another way to develop a more pluralist climate justice framework is to find ways to incorporate various criteria or indicators of justice. These include equality of social positions and powers; equity of rights, resources, and opportunities; human welfare, health, finances, and average life spans; and environmental welfare, including ecosystem services, key species, habitat loss, and the like.

Emerging out of the environmental justice movement of the 1980s, the climate justice movement is tied directly to the Age of the Anthropocene and to climate change caused by the burning of fossil fuels. Climate change particularly impacts people of poor and minority communities who lack economic and political power and who live in areas most impacted by toxics and water, soil, and air pollution. Additionally, more research is needed about the roots of environmental ethics in Eastern and developing countries. How can those ethics and approaches to sustaining the environment become a critical part of saving the earth for the future? The way out of the Anthropocene is through a worldwide movement toward multiple systems of sustainability and the substitution of renewable forms of energy for the burning of fossil fuels. With new forms of ethics and justice, we may move out of the Anthropocene and into a new Age of Sustainability.

EPILOGUE

# The Future of Humanity and the Earth

The Anthropocene raises significant issues not only for the sciences and social sciences but especially for the environmental humanities. Humanity's relationships with the environment should address the consequences of climate change and intersect with ethical and climate justice frameworks to assist vulnerable populations and influence policy and individual choices for the human future. Dichotomies such as nature/culture, ethics/environment, and mind/body are challenged by climate concerns. The humanities have contributed insights to ecological management strategies in ways that respond to and help curtail climate change effects.[1] There are significant linkages and overlapping issues between the focal points of the humanities—art, literature, religion, philosophy, ethics, and justice—that help to build a framework for the resolution of environmental problems facing humanity in the twenty-first century and beyond.

## A New Story

I believe that we need a new story and a new ethic for the twenty-first century, as we are in danger of experiencing another "death of nature" that may include the human species and much of the physical and biological world as it exists today. If in fact we can create that new story of sustainability, we can exit the Age of the Anthropocene.

The new story of sustainability is a framework for a new Age of

EPILOGUE

Sustainability in which humans and the earth are in dynamic interaction and there is a give-and-take between humans and nonhuman nature. It recognizes that nature is autonomous and sometimes unpredictable—a nature described by mechanistic science and also by chaos and complexity theories. As humans, we can learn from what is now happening to the oceans and atmosphere as a result of the anthropogenic accumulation of greenhouse gases that is disrupting life as we know it today. We can use our knowledge of science, technology, and society, along with our spiritual and ethical relations with each other and the nonhuman world, to create a new story for the earth's future.

The story of sustainability is rooted in the idea that humans take from the earth what they need for subsistence, give back what can be regrown and recycled, and leave nonrenewable resources (specifically fossil fuels) within the earth to the extent possible. My use of the term *sustainability,* however, should be distinguished from "sustainable development" as enunciated by Gro Harlem Brundtland in *Our Common Future*—also known as the Brundtland Report (World Commission on Environment and Development 1987).[2] As I have elaborated elsewhere: "Rather than sustainable development, which reinforces dominant approaches to development, women's environmental groups, and many other NGOs, have substituted the term 'sustainable livelihood.' Sustainable livelihood is a people-oriented approach that emphasizes the fulfillment of basic needs, health, employment, and old-age security, the elimination of poverty, and women's control over their own bodies, methods of contraception, and resources. Such approaches are exemplified by localized sustainable agriculture, bioregionalism, and indigenous approaches to sustainability."[3] They include ecological methods that incorporate the wisdom of indigenous peoples and new forms of ecological management and restoration ecology that give back what is taken from the land.

# EPILOGUE

As discussed in chapter 6, "Ethics and Justice," the new ethic that accompanies the new Age of Sustainability is a partnership ethic. It states: *The greatest good for the human and nonhuman communities is in their mutual living interdependence.*

My partnership ethic has five precepts:

1. Equity between human and nonhuman communities.
2. Moral consideration for both humans and other species.
3. Respect for both cultural diversity and biodiversity.
4. Inclusion of women, minorities, and nonhuman nature in the code of ethical accountability.
5. An ecologically sound management that is consistent with the continued health of both the human and the nonhuman communities.[4]

A partnership ethic is based on an exchange among humans and between humans and nature. In other writings, I have provided numerous examples of how to put this ethic into place. I have included ways to work with the business community and within current structures of capitalism, while arguing that a sustainable system must move away from the overexploitation of resources for the sake of profit. Implementing a partnership ethic is critical to the new story of sustainability as an alternative to the negative aspects of the Age of the Anthropocene.

How can sustainability be implemented in order to achieve a better future? According to Mark Jacobson, professor of civil and environmental engineering at Stanford University and director of its Atmosphere/Energy Program, it is possible to implement changes that move us toward sustainability by the year 2050 using wind, water, and solar energies.[5] To achieve a solution to global warming by 2050,

# EPILOGUE

*Figure E.1. Mark Jacobson (b. 1965)*

Jacobson argues, we must move from a fossil fuel–based economy to a renewable energy economy. We must move from the reliance on coal, oil, and gas (COG) to a new economy based on wind, water, and solar (WWS) energy, with the inclusion of 1.2 percent geothermal, tidal, and wave energy. Jacobson points out that WWS is expanding rapidly because it is sustainable, clean, safe, and widely available. The biggest problem is that of grid reliability. He and his colleagues produced a model of a grid that would supply WWS power (with no natural gas, biofuels, or nuclear power) at a reasonable cost and with no grid overloads. The long-term goal is finding a way to provide time-dependent load reliability at low cost combined with storage and demand response. If this could be accomplished, a 100 percent WWS world could exist by 2050.

How might a long-term WWS economy be achieved? Jacobson directs the Solutions Project (TheSolutionsProject.org). He daily uploads new accomplishments and goals around the world that could help to reduce global warming by 2050 through the use of WWS.

# EPILOGUE

Examples of solutions include:

- A total of 139 countries could transition to 100 percent renewable energy under the new plan.
- A full 82 percent of twenty-six thousand respondents in thirteen countries want 100 percent renewables; only 18 percent don't.
- A total of fifty cities and towns across the United States have now committed to transition to 100 percent clean, renewable sources of energy, such as wind and solar power.
- Google is officially 100 percent sun and wind powered—three gigawatts' worth.
- The power grid in South Australia now includes a huge Tesla battery tied to a wind farm, allowing the system to supply electricity around the clock.
- Wind power prices have plummeted again in Germany. The price of onshore wind in Germany is now half the EU's projections for 2030.
- The European Union will hike its renewables goal for 2030 in response to the falling cost of renewables. It is now affordable to raise the current 27 percent target to 30 percent.
- A coal plant burning thirteen thousand tons of coal per day in Kenosha County, Wisconsin, producing one gigawatt of energy, is scheduled to be shut down.
- In Mexico, the renewable energy subsidiary Enel Rinnovabile has been awarded the right to supply energy and clean certificates with four wind projects for a total capacity of 593 megawatts in the country's third long-term public tender.

In June 2017, however, Christopher Clack at the National Oceanic and Atmospheric Administration (NOAA) in Boulder, Colorado, and

# EPILOGUE

*Figure E.2. Christopher Clack*

others responded to Jacobson with an article arguing that we could not achieve a solution without also including nuclear energy and bioenergy.[6] Clack and colleagues stated that to achieve a decarbonized reliable energy system, we need a diverse portfolio of clean energy technologies that is broader than wind, water, and solar alone. This goal cannot be achieved without the use of nuclear and bioenergy sources coupled with carbon capture storage. There are particular industries that at this time are extremely difficult to convert to electric energy, such as the aviation and cement industries. The maximum grid efficiency that can be achieved when these industries are included is 80 percent. My own belief is that, despite the analysis of Clack and his colleagues, building new nuclear energy plants is not only too expensive and therefore not feasible, but also too dangerous to be pursued. A major problem with biofuels, as Paul Crutzen et al. argued in 2007, is that the emission of nitrous oxide ($N_2O$) in the production of biofuels contributes more to global warming than the fossil fuels they replace.[7] Carbon capture through planting of trees

## The Global Ecological Revolution

and plants combined with renewable forms of energy could instead provide a way forward.

We need a future based on sustainable energy that can replace the Age of the Anthropocene as the dominant paradigm for the twenty-first century. To accomplish this change will require not only a transformation to renewable energy but a global social and economic revolution. For the Age of the Anthropocene to be replaced by the Age of Sustainability, we need to alter the capitalist relations of production that constitute the Capitalocene. Moreover, the current age of patriarchy would have to give way to new socioeconomic forms, new gender relationships (male, female, bisexual, and transsexual), and an ecological ethic of partnership with each other and the earth.[8] The relations between ecology, production, reproduction, and consciousness displayed in the diagram here would all have to be transformed. Such a transformation would constitute a global ecological revolution that could lead to a sustainable world. This would resolve problems existing at the intersections of all three levels shown in the diagram: (1) production and ecology, (2) human and nonhuman reproduction, and (3) consciousness.

Problems at the intersection of capitalist economic production and ecology (level 1) include global resource depletion and pollution. Such issues comprise nuclear war along with nuclear power plant accidents, threatening the earth with radioactive cancer-causing emissions. The burning of fossil fuels for industrial production increases carbon dioxide in the atmosphere. The cutting of tropical rain forests for grazing and crops reduces photosynthetic conversion to oxygen, resulting in global warming and melting icecaps. The "greenhouse effect" alters weather patterns that affect agriculture, fishing, and the

# EPILOGUE

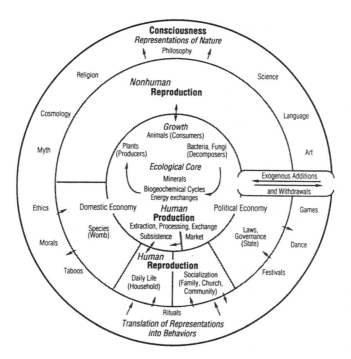

*Figure E.3. Ecological Revolutions diagram*

ecology of local habitats. Nonbiodegradable industrial plastics pollute soils and oceans. As chlorofluorocarbons are produced for refrigerants and Styrofoam packaging, the earth's protective ozone layer is threatened. Toxic wastes from chemical industries enter groundwater supplies, threatening human health. Acid rain from coal-burning "smokestack" industries cross national boundaries, increasing the acidity of lakes and damaging forests. Habitat destruction from industrial expansion endangers hundreds of indigenous species over the whole globe.

Other problems and disjunctions are occurring at the intersection of production and reproduction (level 2). Global population continues to grow exponentially despite declining reproductive rates in developed nations. Increased populations in developing countries put pressure on local economies and consequently on the land. Such pressures challenge traditional sex/gender roles and create new patterns in both industrial/economic production and biological reproduction. The emergence of worldwide "green" political parties is in part a response to the failure of the legal political frameworks that reproduce capitalist society to regulate pollution and depletion. These tensions within production and reproduction are experienced as threats to the health and survival of both human and nonhuman nature.

At level 3, that of consciousness, the mechanistic worldview that goes back to the Scientific Revolution of the seventeenth century and the work of Isaac Newton produced a scientific framework that allowed for predictability, management, and control over the natural world. This framework, when enhanced by technologies of the Anthropocene, such as the steam engine, has made possible great improvements and comforts to human life, but at the expense of deterioration within the natural world. What possibilities exist that could help to resolve the current ecological crisis?

The outcome of the global ecological crisis in production, reproduction, and consciousness could be negative or positive. A pessimistic scenario would be the crisis and collapse predicted by the "limits to growth" models of the 1970s and the Malthusian dilemma of exponential population growth outrunning the food supply. But a positive outcome could be that the crisis and reorganization, implied by the "order out of chaos" approaches of Ilya Prigogine and Erich Jantsch, could move the entire globe toward ecological and economic sustainability in the twenty-first century.[9] New forms of production, reproduction, and consciousness could structure the world differently for twenty-second-century citizens.

The transition to an Age of Sustainability would entail changes

# EPILOGUE

in production and reproduction that emphasize sustainable development in both developed and developing countries. The exploitation of nature and of indigenous and third world peoples would give way to priorities that fulfill subsistence and quality of life needs. These would be enhanced by global efforts to conserve energy and renewable natural resources, recycle nonrenewable resources, and adopt appropriate technologies. If sensitively structured, ecological and economic development in the developing countries could pave the way to the demographic transition that has lowered reproductive rates in developed countries. Changes in production would thus support changes in reproduction and both together would alleviate human pressures on the global ecosystem. This transition would be legitimated by changes in values and in ways that people perceive, know, and experience reality.

Supporting the emergence of a transformation of consciousness are calls by physicists, ecologists, feminists, poets, and philosophers for philosophical changes that would reintegrate culture with nature, mind with body, and male with female modes of experiencing and representing "reality." They suggest that nature as actor may now be breaking out of the mechanistic straitjacket in which human representations have confined it for the past three hundred years. Through the social construction of a new "reality," future generations may learn a worldview that is nonmechanistic. When philosopher Max Horkheimer called for the revolt of nature in 1947, he invited it to speak in a language other than instrumentalism. "Once it was the endeavor of art, literature, and philosophy to express the meaning of things and of life, to be the voice of all that is dumb, to endow nature with an organ for making known her sufferings, or we might say, to call reality by its rightful name. Today nature's tongue is taken away. Once it was thought that each utterance, word, cry, or gesture had an intrinsic meaning; today it is merely an occurrence." The voice with which nature speaks and is heard by humans is tactile, sensual, auditory, odoriferous, and visual—not disembodied reason, but visceral understanding.[10] To survive we must once again actively make ourselves more "like" the environment, not as object, but in the deepest sense of visceral remerging with the earth.

Arising from concerns over the earth's future is a spectrum of new sciences infused with an ecological perspective. At their root is a new form of thinking—integrative thinking. Imitation, synthesis, and a creative reciprocity between humans and nonhuman nature constitute a form of consciousness in which tacit knowing through the body and information networks ("mind") in nature link humans to the nonhuman world. The new theoretical frameworks challenge positivist epistemology through participatory forms of consciousness. Gregory Bateson's "ecology of mind" sees nature as a network of information cycling from brain to hand to stick to rock to earth to eye to brain. "Mind" in nature integrates human subject and active object into a larger network of energy and information exchange. Nature is a changing whole consisting of interactions and processes interpreted by humans. The body's tacit knowledge is one with the mind.[11]

Philosophers have also proposed alternatives to the mechanistic framework based on nature's inherent activity, self-organization, permeable boundaries, and resilience. Deep ecologists, such as Arne Naess (1912–2009), argue that reform environmentalism is insufficient to deal with the magnitude of global environmental problems. They call for a fundamental transformation in Western epistemology, ontology, and ethics. Deep ecology represents a change from a mechanistic to an ecological consciousness rooted in biospecies equality, appropriate technologies, and recycling. Social ecology, as advocated by Murray Bookchin (1921–2006), focuses on bioregions as ecological homes and as sites for local social transformations. The new philosophy is infused with an environmental ethic oriented toward establishing sustainable relations with nature.[12]

# EPILOGUE

Structural changes within science itself may also be indicative of the emergence of a new paradigm. The new physics of David Bohm contrasts the older world picture of atomic fragmentation with a new philosophy of wholeness expressed in the unfolding and enfolding of moments within a "holomovement." His cosmology is one of the primacy of process rather than the domination of parts. The Gaia hypothesis of British chemist James Lovelock proposes that the earth's biota as a whole maintain an optimal chemical composition within the atmosphere and oceans that support its life. Gaia, the name of the Greek earth goddess, is a metaphor for a self-regulating (cybernetic) system that controls the functioning of the earth's chemical cycles. Chaos theory in mathematics offers tools for describing complexity and turbulence consistent with the idea that nature as actor offers surprises and catastrophes that cannot be predicted by linear equations and mechanistic descriptions.[13]

Coupled with these changes in science, epistemology, and ethics are new applied sciences oriented toward effecting a transition to ecological sustainability. Restoration is the active reconstruction of pristine ecosystems (such as prairies, grasslands, rivers, and lakes). By studying and mimicking natural patterns the wisdom inherent in evolution can be re-created. Rather than deconstructing nature and simplifying ecosystems, as the past three centuries of mechanistic science have taught us to do supremely well, restoration ecologists are putting them back together. Rather than analyzing nature for the sake of dominating and controlling it, restorationists are synthesizing it for the sake of living symbiotically within the whole.[14]

A global ecological revolution would reconstruct relations among people and between people and nature. The domination of women and nature inherent in the market economy's use of both as resources would be restructured. Women and men would become partners with nonhuman nature.

## EPILOGUE

An ecological transformation in the deepest sense entails changes in ecology, production, reproduction, and forms of consciousness. Ecology as a new worldview could help resolve environmental problems rooted in the industrial-mechanistic mode of representing nature. In opposition to the subject/object, mind/body, and nature/culture dichotomies of mechanistic science, ecological consciousness sees complexity and process as including both nature and culture. In the ecological model, humans are neither helpless victims nor arrogant dominators of nature, but active participants in the destiny of the webs of which they are a part.

Although many changes leading to a healthier, sustainable biosphere seem to be occurring, the forces that encourage the current patterns of global resource depletion and pollution are still strong. Patriarchy, capitalism, and the domination of nature are deeply entrenched and function to maintain the present direction of development. Yet one may hope that a sustainable global environment, society, and ethic will emerge in the twenty-first century.

Were we able to achieve sustainability, or at least see a turndown in global warming by 2050, we would be exiting the Age of the Anthropocene and entering a new age, that of sustainability, in which humans interact with the earth in partnership, summed up by the following mantra:

> Solar panels on every roof;
> Bicycles in every garage;
> and
> Vegetables in every backyard.[15]

Policies, ethics, and individual actions can indeed restore, reclaim, and reinvigorate the earth.

# Notes

### Preface
1. Crutzen and Stoermer, "Anthropocene."
2. Sörlin, "Environmental Turn in the Human Sciences" and "The Anthropocene," 12.

### Introduction
Portions of this chapter are modified from Wells and Merchant, "Melting Ice."
1. Crutzen and Stoermer, "Anthropocene."
2. Crutzen and Stoermer, "Anthropocene." On James Watt's steam engine, see "engraving of a 1784 steam engine designed by Boulton and Watt," "James Watt," Wikipedia, https://en.wikipedia.org/wiki/James_Watt.
3. Crutzen and Stoermer, "Anthropocene": "Several climatically important 'greenhouse gases' have substantially increased in the atmosphere: $CO_2$ by more than 30% and $CH_4$ by even more than 100%," p. 17
4. Crutzen and Stoermer, "Anthropocene."
5. Jenkins, "Carbon Capture," 32. Also "Global Climate Report."
6. Zalasiewicz et al., "The New World of the Anthropocene": "The ultimate effect on the biosphere of climate change coupled with other human stressors (habitat fragmentation, invasive species, predation) is a sharp increase *in the rate of extinctions.* ... This current human-driven wave of extinctions looks set to become Earth's sixth great extinction event" (2229; italics added). For graphs of Earth System Trends, see Steffen et al., 2004; Bonneuil and Fressoz, *The Shock of the Anthropocene* 10–11. See also Grooten and Almond, *Living Planet Report*, 24, 25: "Large Dams" and "Carbon Dioxide."
7. U.S. Environmental Protection Agency, "Climate Change Science." See https://350.org/. Adrian Raftery and colleagues, using IPCC data from 2013, argue

that "the likely range of global temperature increase is 2.0–4.9°C with a median of 3.2°C and a 5% (1%) chance that it will be less than 2°C (1.5°C)." See Raftery et al., "Less Than 2°C Warming by 2100 Unlikely."

8. Hanania, Stenhouse, and Donev, "Discovery of the Greenhouse Effect"; Enzler, "History of the Greenhouse Effect."

9. Enzler, "History of the Greenhouse Effect"; Hanania, Stenhouse, and Donev, "Discovery of the Greenhouse Effect"; Weart, *The Discovery of Global Warming: A History.*

10. Steffen et al., *Global Change and the Earth System;* Steffen, Crutzen, and McNeill, "The Anthropocene."

11. Enzler, "History of the Greenhouse Effect." See also Maslin, *Global Warming;* www.coolearth.org/IPCC/Global_Warming; http://www.edf.org/climate_change.

12. Eustachewich, "Terrifying Climate Change Warning"; see also Miller and Croft, "Life-or-Death Warning." In the United States, the Fourth National Climate Assessment, which appeared in November 2018, argued that the earth will experience more "hurricanes, tornadoes, floods, fires, water shortages, heat waves, and drought across the globe unless governments take action to stop the release of greenhouse gasses and halt the relentless heating of the earth," Fimrite, "Dire New Forecast on Global Warming."

13. McKibben, *The End of Nature.* See also McKibben, "Can Anyone Stop It?"

14. Homero et al., "The Earth Charter."

15. Union of Concerned Scientists, "Capping Global Warming Emissions"; Keaten, "Melting Ice Opens Route through Arctic."

16. Keaten, "Melting Ice Opens Route through Arctic"; Alexander, "Greenhouse-Gas Emissions Soar."

17. Shepard, "Can't We Just Remove Carbon Dioxide?"; "Ocean Acidification."

18. Leiserowitz, "American Risk Perceptions"; Wike, "What the World Thinks about Climate Change."

19. Lomborg, *Cool It;* Nordhaus and Shellenberger, *Break Through.*

20. Leiserowitz, "American Risk Perceptions."

21. Ellis et al., "Dating the Anthropocene"; Ruddiman, "The Anthropogenic Greenhouse"; Ruddiman, "Early Anthropogenic Hypothesis"; Ruddiman, *Earth Transformed;* Ruddiman, "How Did Humans First Alter Global Climate?"; Crosby, *The Columbian Exchange;* Diamond, *Guns, Germs, and Steel.*

22. Crutzen, "Geology of Mankind"; Steffen, Crutzen, and McNeill, "The An-

thropocene"; McNeill and Engelke, *The Great Acceleration*. On Paul Crutzen's visit to Max Planck Institute for Chemistry in 2012, see Voosen, "Scientists Drive Golden Spike." On the Anthropocene and nuclear winter, see Crutzen, Lax, and Reinhardt, "Paul Crutzen on the Ozone Hole"; and Waters et al., "Nuclear Weapons Fallout."

23. Numerous scholars who have written on the Anthropocene have proposed new names and new starting points for the era. See Haraway, "Anthropocene, Capitaloscene, Plantationocene, and Chthulucene"; Haraway, *Staying with the Trouble*; Moore, *Anthropocene or Capitalocene?*; Moore, "The Capitalocene," parts 1 and 2; Steffen, Crutzen, and McNeill, "The Anthropocene."

24. Samways, "Translocating Fauna to Foreign Lands"; Cox, "Alien Species in North America and Hawaii"; Curnutt, "A Guide to the Homogenocene"; Mann, *1493*, 23, 300, 32, 42, 54 (quotation), 95. On the differences between Homogenocene and Anthropocene, see Mentz, "Anthropocene v. Homogenocene."

25. Wilcox, "Resisting the Plantationocene"; Haraway, "Anthropocene, Capitaloscene, Plantationocene, and Chthulucene"; Haraway, "Tentacular Thinking."

26. On the Gynocene, see Demos, "Anthropocene, Capitalocene, Gynocene": "Additionally, there's the Gynocene thesis, implying a gender-equalized, even feminist-led, interventionist environmentalism, which locates anthropogenic geological violence as coextensive with patriarchal domination, linking ecocide and gynocide. Contesting the ravages of Anthropos it calls for new models of eco-feminist stewardship, resonating as much with Indigenous reverence for Mother Earth and the multifaceted rights-of-nature mobilizations in South America, as with the post-heteronormative, eco-sexualist care for Earth-as-Lover, as appearing in the carnivalesque Earth-marriage ceremonies of performance artists Beth Stephens and Annie Sprinkle, who deploy matrimony as a radical act against environmental destruction, including mountain-top removal mining in North America," para. 4.

27. Chakrabarty, "The Climate of History."

28. Chakrabarty, "The Climate of History"; see 211 for quote from Davis, "Living on the Ice Shelf."

29. Klein, *This Changes Everything*; Klein, "Radical Guide to the Anthropocene."

30. Angus, "Confronting the Climate Change Crisis"; Angus, *The Global Fight for Climate Justice*; Angus, *Facing the Anthropocene*; Angus, "Does Anthropocene Science Blame All Humanity?"

31. Viveiros de Castro, "Exchanging Perspectives."

32. Moore, *Anthropocene or Capitalocene?*; Moore, *Ecology in the Web of Life*.

33. For example, see Hecht, "The African Anthropocene"; Spangenberg, "China in the Anthropocene"; Totman, *Japan;* Elvin, *The Retreat of the Elephants.*

### Chapter 1. History

1. Crutzen and Stoermer, "Anthropocene."
2. My first book was titled *The Death of Nature: Women, Ecology, and the Scientific Revolution.* In it I argued that the organic worldview of ancient times up to the Renaissance had a body, soul, and spirit and the earth was a nurturing mother. It was replaced in the Scientific Revolution of the seventeenth century by a mechanistic worldview in which matter was dead and God was a clockmaker, an engineer, and a mathematician. This change constituted "the [first] death of nature."
3. On the history of the Enlightenment and its manifestations in many fields, see https://www.newworldencyclopedia.org/entry/Age_of_Enlightenment.
4. Fleming, "Latent Heat and the Invention of the Watt Engine"; https://www.britannica.com/biography/Joseph-Black.
5. https://www.britannica.com/biography/Antoine-Lavoisier.
6. On the operation and history of the Newcomen engine, see http://www.animatedengines.com/newcomon.html; and https://www.egr.msu.edu/~lira/supp/steam/: "The pressure difference between the atmosphere and the resulting vacuum *pushes the steam piston down,* pulling the main pump piston upwards, lifting the water above the main pump piston and filling the lower main pump chamber with water." See also www.sjsu.edu/faculty/watkins/newcomen5.htm: "The way the Newcomen engine worked ... was with a piston in a cylinder connected to a rocker arm attached to a pump. First the cylinder was filled with steam from a boiler. This pushed the piston up. Then water was sprayed into the cylinder creating a vacuum. This pushed the piston down pulling the pump rod on the other side of the rocker arm up, thus lifting the water. The opening and closing of valves for the alternating injection of steam and water was self-actuating so the engine and pump could operate continuously."
7. On James Watt's steam engine, see "Steam Engine."
8. Wilson, "Sadi Carnot," 134.
9. Carnot, *Réflexions sur la puissance motrice du feu;* Carnot, *Reflections on the Motive Power of Fire,* 3–69. With the aid of a large center-pivot lever and wheel, the motion of the piston could be transformed into the up and down motion of lifting coal out of a mine or the lateral motion of pushing or pulling a cart or other vehicle.

10. Carnot, *Reflections on the Motive Power of Fire*, 3; see also Fox, *The Caloric Theory of Gases*.

11. On the term *motive power,* see http://www.eoht.info/page/Motive+power. On *caloric,* see https://www.britannica.com/science/caloric-theory: "Caloric theory, explanation, widely accepted in the 18th century, of the phenomena of heat and combustion in terms of the flow of a hypothetical weightless fluid known as caloric. The idea of an imaginary fluid to represent heat helped explain many but not all aspects of heat phenomena. It was a step toward the present conception of energy—i.e., that it remains constant through many physical processes and transformations; however, the theory also deterred clear scientific thinking. The caloric theory was influential until the mid-19th century, by which time many kinds of experiments, primarily with the mechanical equivalent of heat, forced a general recognition that heat is a form of energy transfer and, in particular, that limitless amounts of heat could be generated by doing work on a substance." See also Fox, *The Caloric Theory of Gases,* 183–89.

12. The term *Carnot cycle* seems to have first been used in 1887 in the *Encyclopaedia Britannica,* 22:481–82. See https://www.britannica.com/science/Carnot-cycle, "Carnot cycle, in heat engines, ideal cyclical sequence of changes of pressures and temperatures of a fluid, such as a gas used in an engine, conceived early in the 19th century by the French engineer Sadi Carnot. It is used as a standard of performance of all heat engines operating between a high and a low temperature." The term was also used by University of Wisconsin, Madison, physicist John C. Shedd in 1899. See Shedd, "A Mechanical Model of the Carnot Engine": "There is, perhaps, no proposition in the range of Physics that is more difficult of comprehension by the average student than that embodies in the so-called Carnot Engine and Carnot Cycle" (174). "Carnot engine" was used by Scottish physicist James Maxwell in 1871; see Maxwell, *Theory of Heat,* 148.

13. On the Carnot cycle, see https://chem.libretexts.org/Core/Physical_and _Theoretical ... Cycles/Carnot_Cycle. On the definition of work, see https://www .thoughtco.com/definition-of-work-in-chemistry-605954.

14. See https://thebiography.us/en/clapeyron-benoit-paul-emile. Also Clapeyron, "Memoire sur la puissance motrice de la chaleur"; Clapeyron, "Memoir on the Motive Power of Heat," 73–74. Merriam-Webster defines the Carnot cycle "as an ideal reversible closed thermodynamic cycle in which the working substance goes through the four successive operations of isothermal expansion to a desired point, adiabatic expansion to a desired point, isothermal compression, and adiabatic com-

pression back to its initial state" (https://www.merriam-webster.com/dictionary/Carnot%20cycle).

15. Clapeyron, "Memoir on the Motive Power of Heat," 74, italics added.

16. Clapeyron, "Memoir on the Motive Power of Heat," 78, 79 (fig. 2). "In the same way it can be shown that no gas or vapor exists which, if used to transmit heat from a hot body to a cold one by the methods described, can develop a quantity of action greater than any other gas or vapor" (81). Also, "There is a loss of *vis viva* whenever there is contact between bodies at different temperatures" (81). Here vis viva is $mv^2$, the measure of force identified by Gottfried Wilhelm Leibniz in 1686 and later, with the addition of one-half, called kinetic energy or $½ mv^2$.

17. Clapeyron, "Memoir on the Motive Power of Heat," 81. See editor Eric Mendoza's note: "This extraordinary paragraph is an unambiguous statement of the First Law of Thermodynamics. It serves to emphasize ... that the caloric theory and the *vis viva* theory were not regarded as mutually exclusive."

18. Clausius, "Über die bewegende Kraft der Wärme und die Gesetze," translated as "On the Moving Force of Heat and the Laws." On Clausius's statement of "Carnot's Principle," see 372 in the German source. On what became known as the second law of thermodynamics, see 501: "Ein Uebergang von Wärme aus einem warmen in einen kalten Körper findet allerdings in solchen Fällen statt, wo Arbeit durch Wärme erzeugt, und zugleich die Bedingung erfüllt wird, dass der wirksame Stoff sich am Schlusse wieder in demselben Zustande befinde, wie zu Anfang." Eng. trans., Clausius, "On the Motive Power of Heat, and on the Laws which can be Deduced from it for the Theory of Heat Itself," in William F. Magie, trans. and ed., *The Second Law of Thermodynamics: Memoirs by Carnot, Clausius, and Thomson* (New York: Harper, 1899), 65–106, see section 2, quotations on 88 and 89. See also Carnot, *Reflections on the Motive Power of Fire*, 132, 133.

19. Clausius, "Über eine veränderte Form des zweiten Hauptsatzes der mechanischen Wärmetheorie," translated as "On a Modified Form of the Second Fundamental Theorem in the Mechanical Theory of Heat," quotation on 86. On the first page of this paper, in a footnote, Clausius states, "The present memoir appeared in Poggendorff's *Annalen*, vol. xciii, p. 481, and was referred to by the author in a letter lately published in this Magazine; it is also employed to a considerable extent in a memoir on the steam-engine by the same author, a translation of which will shortly appear." Clausius began, "In my memoir 'On the Moving Force of Heat, and the Laws which can be deduced therefrom,' I have shown that the theorem of the equiva-

lence of heat and work, and Carnot's theorem, are not mutually exclusive, but that, by a small modification of the latter, which does not affect its principal part, they can be brought into accordance." See also https://www.britannica.com/biography/Rudolf-Clausius.

It should be noted that Clausius's statements of what came to be known as the second law of thermodynamics in the 1850 (English 1851) and 1856 papers are written differently. Confusion has existed in the secondary literature about the two different statements and the two different citations.

20. Clausius, "Über vershiedene für Anwendung bequeme Formen der Hauptgleichungen der mechanishen Wärmetheorie," republished as Clausius, *Mechanical Theory of Heat*, Ninth Memoir, 327–65.

21. Clausius, "On Several Convenient Forms of the Fundamental Equations of the Mechanical Theory of Heat," in *Mechanical Theory of Heat*, Ninth Memoir, 357: "I propose to call the magnitude S the entropy of the body, from the Greek word [ἐντροπία], transformation. I have intentionally formed the word entropy so as to be as similar as possible to the word energy; for the two magnitudes to be denoted by these words are so nearly allied in their physical meanings, that a certain similarity in designation appears to be desirable." See also Clausius, "Application of the Two Fundamental Theorems of the Mechanical Theory of Heat to the Entire Condition of the Universe," in *Mechanical Theory of Heat*, 365: "We may express in the following manner the fundamental laws of the universe which correspond to the two fundamental theorems of the mechanical theory of heat. 1. *The energy of the universe is constant.* 2. *The entropy of the universe tends to a maximum.*" Clausius, "On a Mechanical Theorem Applicable to Heat"; also Magie, *Source Book in Physics*, 234 (on entropy), 236 (quotation). On Carnot and Clausius, see Mach, *Principles of the Theory of Heat*; Hiebert, *Historical Roots of the Principle of the Conservation of Energy*; Newburgh, "Carnot to Clausius."

22. "Entropy and Heat Death."

23. See https://www.britannica.com/biography/William-Thomson-Baron-Kelvin.

24. Kelvin, "On the Dynamical Theory of Heat," 8, 13. In a footnote, he wrote: "If this axiom be denied for all temperatures, it would have to be admitted that a self-acting machine might be set to work and produce mechanical effect by cooling the sea or earth, with no limit but the total loss of heat from the earth and sea, or, in reality, from the whole material world" (13). See also Thomson, *Mathematical and Physical Papers*, 175.

25. Joule, "On Changes of Temperature," 381.

26. See https://www.britannica.com/biography/James-Prescott-Joule.

27. Michael Fowler, University of Virginia, Spring 2002, see http://galileo.phys.virginia.edu/classes/152.mf1i.spring02/Joule.htm. "Joule also calculated that the water just beyond the bottom of a waterfall will be one degree Fahrenheit warmer than the water at the top for every 800 feet of drop, approximately, the kinetic energy turning to heat as the water crashed into rocks at the bottom. Joule spent his honeymoon at Chamonix in the French Alps, and Lord Kelvin claimed later that when he chanced to meet the honeymooners in Switzerland, Joule was armed with a large thermometer to check out the local waterfalls (but it is generally believed that Kelvin made this up)."

28. See https://www.physlink.com/Education/AskExperts/ae181.cfm. See also https://www.wolframscience.com/reference/notes/1019b.

29. Kelvin, "On the Universal Tendency in Nature to the Dissipation of Mechanical Energy." See also Smith and Wise, *Energy and Empire*, 500–501.

30. Kelvin, "On the Universal Tendency in Nature to the Dissipation of Mechanical Energy," 304–6, italics added.

31. Thomson, *Mathematical and Physical Papers*, 232.

32. "Entropy and Heat Death."

33. See "What Exactly Is the Heat Death of the Universe?" See also https://phys.org/news/2015-09-fate-universeheat-death-big-rip.html.

34. Rankine, *Manual of the Steam Engine*. On thermodynamics, see "Principles of Thermodynamics," 299–310. See also Rankine, "On the Mechanical Action of Heat"; Rankine, "On the General Law of the Transformation of Energy."

35. On Ludwig Boltzmann and his entropy formula for the kinetic theory of ideal gases, see https://www.britannica.com/biography/Ludwig-Boltzmann and http://www.eoht.info/page/S+%3D+k+ln+W. In statistical mechanics, Boltzmann's equation $S = k.\log W$ is a probability equation relating the entropy $S$ of an ideal gas to the quantity $W$, the number of real microstates corresponding to the gas's macrostate where $k_B$ is the Boltzmann constant (also written as simply $k$) and equal to $1.38065 \times 10^{-23}$ J/K (Joules per Kelvin). In short, the Boltzmann formula shows the relationship between entropy and the number of ways the atoms or molecules of a thermodynamic system can be arranged. I thank Persi Diaconis of Stanford University for bringing this equation to my attention.

36. Prigogine, "Time, Structure, and Fluctuations."

37. Prigogine and Stengers, *Order out of Chaos*. The above two paragraphs rely on Merchant, *Reinventing Eden*. On the change from the perception held during most of early history and as described mathematically by Ptolemy (100–160 BCE) to the sun-centered universe of Nicolaus Copernicus (1473–1543) in 1543, see Kuhn, *The Copernican Revolution;* and Kuhn, *The Structure of Scientific Revolutions*.

## Chapter 2. Art

1. Quoted in Jackman, *The Development of Transportation in Modern England*, 2:497–98, quotation on 498.
2. Jackman, *The Development of Transportation in Modern England*, 2:497.
3. Howarth, "Ten Turner Paintings."
4. Thomas, "The Chase."
5. On Édouard Manet's railroad paintings, see https://www.theguardian.com/uk/2005/apr/14/transport.
6. On Claude Monet's railroad paintings, see https://www.theguardian.com/uk/2005/apr/14/transport.
7. See Cronon, http://www.williamcronon.net/courses/469/handouts/469-telling-tales-on-canvas.html. See also Cronon, "Telling Tales on Canvass."
8. Merchant, *Reinventing Eden*, 109.
9. Cronon, "Telling Tales on Canvass," 85, 84 (fig. 51).
10. On black labor in the railroad industry, see https://blackthen.com/the-four-major-rail-networks-enslaved-african-labor-help-build-in-north-america/; usatoday30.usatoday.com/money/general/2002/02/21/slave-railroads.htm; and https://opinionator.blogs.nytimes.com/2012/02/10/been-workin-on-the-railroad/. On women's labor in railroads, see, http://www.interrail-signal.com/women-workin-on-the-railroad/. On black railroad engineers, see http://www.allenandallenmodelrailroading.com/Rail-History.html. See also Simon, "Railroad Paintings and Art." On women working on railroads, see https://www.pinterest.com/gsrm/women-in-railroading/. On women executives in railroad companies, see the website of the League of Railway Women (LRW), http://lriw.org/.
11. Thorne, *The Singularity Is Coming . . . !*
12. Eliasson, *Your Mobile Expectations*.
13. Cape Farewell Project, "The Art of Climate Change."
14. Morrison, "Envisioning Change."

## Chapter 3. Literature

1. Gulliford, "Love Stories"; Dillard, *Pilgrim;* Kingsolver, *Flight Behavior.*
2. Wordsworth, "The Excursion," 1814, in *Collected Poems,* 1037. For an analysis of Wordsworth's negative views of the steam engine, see Schwartz, "The Industrial Revolution and the Railroad System," "Opposing Voices." On literature in the Anthropocene, see François, "Ungiving Time."
3. Wordsworth, "Steamboats, Viaducts, and Railways," 1833, in *Collected Poems,* 569.
4. Wordsworth, "On the Projected Kendal and Windermere Railway," 1844, in *Collected Poems,* 336. On Wordsworth's and others' support for Thirlmere in the Lake District, see Ritvo, "Fighting for Thirlmere."
5. Wordsworth, in *Collected Poems,* 337.
6. Dickens, *Dombey and Son,* 67–68. See Baumgarten, "Railway/Reading/Time"; Mullan, "Railways in Victorian Fiction."
7. Dickens, *Dombey and Son,* 68. See Arac, "The House and the Railroad."
8. Dickens, *Dombey and Son,* 68. See Mullan, "My Favorite Dickens."
9. Dickens, *Dombey and Son,* 68.
10. Dickens, *Dombey and Son,* 298–99.
11. Dickens, *Hard Times,* ch. 17.
12. Dickens, *Hard Times,* ch. 17.
13. Hawthorne, *The House of the Seven Gables.*
14. Hawthorne, *The Celestial Railroad.*
15. Hawthorne, *Celestial Railroad.*
16. Hawthorne, *Celestial Railroad.*
17. Hawthorne, *Celestial Railroad.*
18. Marx, *The Machine in the Garden.*
19. Emerson, "The Young American." See also https://www.britannica.com/biography/Ralph-Waldo-Emerson.
20. Whitman, "To a Locomotive in Winter." On Greg Bartholomew's musical composition and the performance of Whitman's poem by the Seattle Pro Musica, see http://www.gregbartholomew.com/locomotive.html.
21. Dickinson, "The Railway Train," in *Poems of Emily Dickinson,* poem 17.
22. Frost, "A Passing Glimpse," in *West-Running Brook.* See http://literature.oxfordre.com/view/10.1093/acrefore/9780190201098.001.0001/acrefore-9780190201098-e-635.
23. Snyder, *Riprap and Cold Mountain Poems,* 23. Copyright © 1958, 1959, 1965

by Gary Snyder, from *Riprap and Cold Mountain Poems*. Reprinted by permission of Counterpoint Press.

24. Dillard, *Pilgrim at Tinker Creek,* quotations from ch. 10.

25. McPhee, "Coal Train."

26. McPhee, "Coal Train," part 2.

27. Ghosh, *The Great Derangement,* 73–75, 162, quotations on 75; Ghosh, *The Circle of Reason.*

28. Lo, "How Fast Will Jet Fuel Consumption Rise?"

29. Lo, "How Fast Will Jet Fuel Consumption Rise?" quotations from abstract and p. 3; "Jet Fuel Consumption by Country," https://www.indexmundi.com/energy/?product=jet-fuel. See also Total Fuel Consumption of U.S. Airlines from 2004 to 2017 (in billions of gallons).

30. Patel, "Airplanes Flying on Biofuels."

31. Kingsolver, *Flight Behavior;* Martyris, "Barbara Kingsolver, Barack Obama, and the Monarch Butterfly."

32. Waldman, "Anthropocene Blues."

33. Solnick, *Poetry and the Anthropocene.*

34. Trexler, *Anthropocene Fictions.*

35. Association for the Study of Literature and the Environment, 2016, https://www.asle.org/calls-for-contributions/c21-special-issue-literature-anthropocene/.

36. The term *Phallocene* is used by LaDanta LasCanta (2017), a Venezuelan ecofeminist group.

37. Pereira Savi, "The Anthropocene." See also Grusin, *Anthropocene Feminism.*

38. Bennett, *Vibrant Matter,* viii.

39. Pereira Savi, "The Anthropocene," concluding paragraph.

40. Stevens, Tait, and Varney, *Feminist Ecologies;* Macilenti, *Characterising the Anthropocene;* Major, *Welcome to the Anthropocene.*

## Chapter 4. Religion

1. Grim and Tucker, *Ecology and Religion,* introduction and ch. 5, "Emergence of the Field of Religion and Ecology."

2. White, "Historical Roots of Our Ecologic Crisis," 1205.

3. Merchant, *The Death of Nature.*

4. For essays on religion and the Anthropocene, see Deane-Drummond, Bergmann, and Vogt, *Religion in the Anthropocene.*

5. Merchant, *Reinventing Eden.*

6. Allison, *Religious Organizations Taking Action on Climate Change*.
7. See "Healing Our Planet Earth (HOPE) Conference."
8. Interfaith Center on Corporate Responsibility, "Priorities of ICCR."
9. "Climate Change." Many other evangelical groups, however, deny the reality of climate change.
10. Tucker and Grim, *Religions of the World and Ecology;* Grim and Tucker, *Ecology and Religion.*
11. Tucker, "The Emerging Alliance of Religions and Ecology."
12. Taylor, *Encyclopedia of Religion and Nature.*
13. Macintyre, "Pope to Make Climate Action a Moral Obligation."
14. Pope Francis, "Laudato Si."
15. Faiola, "Pope Francis Presents Trump with a 'Politically Loaded' Gift." See also http://www.hcn.org/issues/49.16/activism-why-religious-communities-are-taking-on-climate-change. Also https://news.mongabay.com/2018/01/popes-message-to-amazonia-inspires-hope-but-will-it-bring-action/.
16. Ruether, *Integrating Ecofeminism, Globalization and World Religions.* Ruether dedicated her book to TREES: The Theoretical Roundtable on Ecological Ethics, at the Graduate Theological Union in Berkeley, California.
17. Goodman, "Native American Activist Winona LaDuke."
18. Allison, "The Spiritual Significance of Glaciers."
19. Spangenberg, "China in the Anthropocene"; Elvin, *The Retreat of the Elephants;* Totman, *Japan;* Ghosh, *The Great Derangement,* esp. 96–98, 103–8, 149, 159–62.
20. From Merchant, *Radical Ecology,* 107–8.
21. Needham, *Science and Civilization in China,* vol. 2; Callicott and Ames, *Nature in Asian Traditions of Thought;* Elvin, *The Retreat of the Elephants;* Totman, *Japan.*
22. Lao Tzu, *The Tao-Teh King,* ch. 51; Capra, *The Tao of Physics.*
23. Grim and Tucker, *Ecology and Religion,* 121–25.
24. This portion of the chapter has been modified from Merchant, *Radical Ecology,* 133–36. See Cobb, "Ecology, Science, and Religion," 99–113, esp. 99, 107–8.
25. Cobb and Griffin, *Process Theology* 79, quotations 76, 155.
26. Griffin, *Unprecedented.*
27. McDaniel, "Physical Matter"; McDaniel, "Christian Spirituality"; McDaniel, *Of God and Pelicans.*
28. McDaniel, "Process Philosophy," unpublished paper.

29. McDaniel, "Process Philosophy."
30. Watts, "'For Us the Land Is Sacred.'"
31. Tory, "Religious Communities Are Taking on Climate Change."
32. Aberra, "The Religious Case for Caring about Climate Change."
33. Bentley, "Muslim Environmentalists."
34. Abraham, "Caring for Creation."
35. *The Earth Charter Initiative.*

## Chapter 5. Philosophy

1. Goldstein, *Plato at the Googleplex.*
2. Goldstein, *Plato at the Googleplex,* 71.
3. Goldstein, *Plato at the Googleplex,* 70.
4. Aristophanes, *The Clouds.*
5. Goldstein, *Plato at the Googleplex,* 98–100.
6. Goldstein, *Plato at the Googleplex,* 105–6, 117.
7. Goldstein, *Plato at the Googleplex,* 119.
8. *Early Greek Philosophy,* 55–67.
9. Nahm, *Early Greek Philosophy,* 84–97, quotation on 91, fragments 41–42.
10. Nahm, *Early Greek Philosophy,* 113–21.
11. Nahm, *Early Greek Philosophy,* 128–48.
12. Nahm, *Early Greek Philosophy,* 149–55, 160–219.
13. Merchant, *The Death of Nature,* 204, 208, 276–78.
14. Nahm, *Early Greek Philosophy,* 68–83.
15. Nahm, *Early Greek Philosophy,* 68. On the discovery of irrational numbers by the Pythagorean Hippasus, see https://www.britannica.com/biography/Hippasus-of-Metapontum.
16. On the background of the Scientific Revolution, see Merchant, *The Death of Nature,* esp. chs. 9, 12.
17. On Leibniz's computer, see Dalakov et al., "The Stepped Reckoner." On binary coding, see https://www.britannica.com/technology/binary-code. On the history of computers, see Yaqoob, "Introduction to Computers, History and Applications."
18. See Merchant, *Autonomous Nature,* 144, 151.
19. Merchant, *Autonomous Nature,* 6, 152.
20. Merchant, *Autonomous Nature,* 152.
21. Merchant, *Autonomous Nature,* 151–52.

22. Waldrop, *Complexity*; Wells, *Complexity and Sustainability*.
23. Botkin, *Discordant Harmonies*, 25.
24. Dara Khosrowshahi, quoted in Said, "Uber Is on the Road," C-3.

## Chapter 6. Ethics and Justice

1. Merchant, *Radical Ecology*, 64.
2. From Merchant, *Radical Ecology*, 64.
3. From Merchant, *Radical Ecology*, 72.
4. From Merchant, *Radical Ecology*, 75, 76.
5. From Merchant, *Radical Ecology*, 81–83.
6. From Merchant, *Radical Ecology*, 83–84. See also Merchant, *Earthcare*, 216–24; Merchant, "Partnership Ethics"; and Merchant, "Partnership with Nature."
7. From Merchant, *Radical Ecology*, 84.
8. IPCC, *Climate Change 2001*, 2.
9. Singer, "Ethics and Climate Change."
10. Gardiner, "A Perfect Moral Storm."
11. Gardiner, "A Perfect Moral Storm," 397.
12. Gardiner, "A Perfect Moral Storm," 397.
13. Russell, "Environmental Racism"; Stewart, "Home May Rise on Incinerator Site."
14. From Merchant, *Radical Ecology*, 170–76.
15. https://www.nrdc.org/flint.
16. See Merchant, *Radical Ecology*, 171–72.
17. Merchant, *Radical Ecology*, 172. Bullard, ed., *Confronting Environmental Racism*, 15–16.
18. Rayner and Malone, *Human Choice and Climate Change*, 219.
19. U.S. Commission on Civil Rights, "Not in My Backyard."
20. United States Global Change Research Program, "National Assessment."
21. Hanna, "Native Communities and Climate Change."
22. Kendall-Miller, "Native American Rights Fund News."
23. Paavola and Adger, *Fairness in Adaptation to Climate Change*.

## Epilogue

Portions of the following chapter are drawn from Merchant, *Ecological Revolutions*.

1. Parker et al., "Climate Change and Pacific Rim Indigenous Nations."

2. Brundtland, *Our Common Future*.
3. Merchant, *Radical Ecology*; see also Braidotti et al., *Women, the Environment, and Sustainability*.
4. Merchant, *Major Problems in American Environmental History*, 224.
5. Jacobson et al., "Low-Cost Solution to the Grid Reliability Problem."
6. Clack et al., "Evaluation of a Proposal for Reliable Low-Cost Grid Power."
7. Crutzen et al., "$N_2O$ Release from Ago-Biofuel Production."
8. On reductionist and community-based ecological approaches, see Worster, *Nature's Economy*, chs. 14, 15.
9. The thermodynamics of Ilya Prigogine contrasts the equilibrium and near-equilibrium dynamics of closed, isolated physical systems described by the mechanistic model with open biological and social systems in which matter and energy are constantly being exchanged with their surroundings. When biological systems are confronted with catastrophic changes, a major reorganization can be triggered. Nonlinear relationships and positive feedbacks support a new development.
10. Horkheimer, *Eclipse of Reason*, 101, 115. On mimesis, see also Berman, *Reenchantment of the World*, 177–82, 69 ff.
11. Bateson, *Mind and Nature*, 237–64.
12. Naess, "The Shallow and the Deep"; Devall and Sessions, *Deep Ecology*; Bookchin, *Ecology of Freedom*.
13. Prigogine and Stengers, *Order out of Chaos*; Jantsch, *Self-Organizing Universe*; Bohm, *Wholeness and the Implicate Order*; Briggs and Peat, *Looking Glass Universe*; Gleick, *Chaos*; Lovelock, *Gaia*.
14. Portions of the following text were originally published as Merchant, "Restoration and Reunion with Nature," 68–70. See also Jordan, "Thoughts on Looking Back," 2; Jordan, "On Ecosystem Doctoring," 2.
15. Carolyn Merchant, "Afterword," in Worthy, Allison, and Bauman, *After the Death of Nature*.

# Bibliography

Aberra, Nesima. "The Religious Case for Caring about Climate Change." *Vox,* April 19, 2017. https://www.vox.com/conversations/2017/4/19/15271166/climate-change-religious-arguments.

Abraham, John. "Caring for Creation Makes the Christian Case for Climate Change." *Guardian,* October 10, 2016. https://www.theguardian.com/environment/climate-consensus-97-per-cent/2016/oct/10/caring-for-creation-makes-the-christian-case-for-climate-action.

Alexander, Kurtis. "Greenhouse-Gas Emissions Soar, Stalling Global Warming Battle." *San Francisco Chronicle,* December 6, 2018, 1, 10.

Allison, Elizabeth. *Religious Organizations Taking Action on Climate Change.* Garrison, NY: Garrison Institute, 2007.

———. "The Spiritual Significance of Glaciers in an Age of Climate Change." *Wiley Interdisciplinary Reviews: Climate Change* 6, no. 5 (2015): 493–508.

Angus, Ian. "Confronting the Climate Change Crisis: An Ecosocialist Perspective," 2008. http://www.readingfromtheleft.com/PDF/ConfrontingTheClimateChangeCrisis2.pdf.

———. "Does Anthropocene Science Blame All Humanity?" http://climateandcapitalism.com/2015/05/31/does-anthropocene-science-blame-all-humanity/.

———. *Facing the Anthropocene: Capitalism and the Crisis of the Earth System.* New York: Monthly Review, 2016.

———, ed. *The Global Fight for Climate Justice: Anticapitalist Responses to Global Warming and Environmental Destruction.* London: Resistance Books, 2009, 2011.

*Anthropocene.* Journal. http://www.elsevier.com/locate/ancene.

Arac, Jonathan. "The House and the Railroad: *Dombey and Sons* and *The House of the Seven Gables*." *New England Quarterly* 51, no. 1 (March 1978): 3–22.

Aridjis, Homero, et al. "The Earth Charter." The Earth Charter Commission, 2000. http://www.earthcharter.org/.

Aristophanes. *The Clouds*. Edited by J. M. Starkey. London: Macmillan, 1911.

Athanasiou, Tom, and Paul Baer. *Dead Heat: Global Justice and Global Warming*. New York: Seven Stories, 2002.

Baer, Paul. "The Worth of an Ice Sheet." Eco-Equity, 2007. www.ecoequity.org.

Bateson, Gregory. *Mind and Nature: A Necessary Unity*. New York: Dutton, 1979.

Baumgarten, Murray. "Railway/Reading/Time: *Dombey & Son* and the Industrial World." *Dickens Studies Annual* 19 (1990): 65–89.

Bennett, Jane. *Vibrant Matter: Political Ecology of Things*. Durham: Duke University Press, 2010.

Bentley, Chris. "Muslim Environmentalists Give Their Religion—and Their Mosques—a Fresh Coat of Green," January 4, 2017. http://capeandislands.org/post/muslim-environmentalists-give-their-religion-and-their-mosques-fresh-coat-green#stream/0.

Berman, Morris. *The Reenchantment of the World*. Ithaca: Cornell University Press, 1981.

Biermann, Frank. *Earth System Governance: World Politics in the Anthropocene*. Cambridge, MA: MIT Press, 2014.

Birch, Charles, and John Cobb Jr. *The Liberation of Life: From the Cell to the Community*. Cambridge: Cambridge University Press, 1981.

Bohm, David. *Wholeness and the Implicate Order*. Abingdon, UK: Taylor & Francis, 2002.

Bonifazi, Conrad. *The Soul of the World: An Account of the Inwardness of Things*. Lanham, MD: University Press of America, 1978.

Bonneuil, Christophe, and Jean-Baptiste Fressoz. *The Shock of the Anthropocene*. London: Verso, 2015.

Bookchin, Murray. *Ecology of Freedom: The Emergence and Dissolution of Hierarchy*. Palo Alto: Cheshire Books, 1982.

# BIBLIOGRAPHY

Botkin, Daniel. *Discordant Harmonies: A New Ecology for the Twenty-First Century.* New York: Oxford University Press, 1990.

Braidotti, Rosi, et al. *Women, the Environment, and Sustainability: Towards a Theoretical Synthesis.* Atlantic Highlands, NJ: Zed Books, 1994.

Briggs, John, and David Peat, *Looking Glass Universe: The Emerging Science of Wholeness.* New York: Simon & Shuster, 1986.

Brown, Donald et al. "White Paper on the Ethical Dimensions of Climate Change." Rock Ethics Institute, University of Pennsylvania, Philadelphia, 2006.

Brown, Peter G., and Peter Timmerman, eds. *Ecological Economics for the Anthropocene: An Emerging Paradigm.* New York: Columbia University Press, 2015.

Brundtland, Gro Harlem. *Our Common Future.* World Commission on Environment and Development. New York: Oxford University Press, 1987.

Buckland, David. Cape Farewell Project, 2005. http://www.capefarewell.com/.

Bullard, Robert D., ed. *Confronting Environmental Racism.* Cambridge: South End Press, 1993.

Callicott, J. Baird, and Roger T. Ames. *Nature in Asian Traditions of Thought.* Albany: State University of New York Press, 1989.

Cape Farewell Project. "The Art of Climate Change," November 5, 2005. http://www.we-make-money-not-art.com/archives/007390.php.

Capra, Fritjof. *The Tao of Physics: An Exploration of the Parallels between Modern Physics and Ancient Mysticism.* Boston: Shambhala, 1991.

Carnot, Sadi. *Reflections on the Motive Power of Fire.* In *Reflections on the Motive Power of Fire by Sadi Carnot and Other Papers on the Second Law of Thermodynamics by E. Claperon and R. Clausius,* edited by Eric Mendoza. New York: Dover, 1960.

———. *Réflexions sur la puissance motrice du feu et sur les machines propres à développer cette puissance.* Paris: Bachelier, 1824.

Chakrabarty, Dipesh. "The Climate of History: Four Theses." *Critical Inquiry* 35, no. 2 (2009): 197–222.

Chamberlin, Roy B., and Herman Feldman. *The Dartmouth Bible: An Abridgment of the King James Version, with Aids to Its Understanding as*

*History and Literature, and as a Source of Religious Experience.* Boston: Houghton Mifflin, 1961.

Clack, Christopher T. M., et al. "Evaluation of a Proposal for Reliable Low-Cost Grid Power with 100% Wind, Water, and Solar." *Proceedings of the National Academy of Sciences* 114, no. 26 (June 27, 2017): 6722–27.

Clapeyron, Émile. "Memoire sur la puissance motrice de la chaleur." *Journal de l'Ecole Royale Polytechnique* 14 (1834): 153–90.

———. "Memoir on the Motive Power of Heat." In *Reflections on the Motive Power of Fire by Sadi Carnot and Other Papers on the Second Law of Thermodynamics by E. Clapeyron and R. Clausius,* edited by Eric Mendoza. New York: Dover, 1960.

Clausius, Rudolf. *The Mechanical Theory of Heat: With Its Applications to the Steam Engine...* London: John Van Voorst, 1867.

———. "On a Mechanical Theorem Applicable to Heat." *Philosophical Magazine,* ser. 4, vol. 40, no. 265 (1870): 122–27.

——— "On a Modified Form of the Second Fundamental Theorem in the Mechanical Theory of Heat." *Philosophical Magazine,* ser. 4, vol. 12, no. 77 (1856): 81–98.

———. "On the Moving Force of Heat and the Laws Regarding the Nature of Heat Itself Which Are Deducible Therefrom." *Philosophical Magazine,* ser. 4, vol. 2, no. 8 (1851): 1–21, 102–19.

———. "Über die bewegende Kraft der Wärme und die Gesetze, welche sich daraus für die Wärmelehre selbst ableiten lassen," *Annalen der Physik und Chemie,* ser. 3, vol. 79 (1850): 368–97, 500–524.

———. "Über eine veränderte Form des zweiten Hauptsatzes der mechanischen Wärmetheorie," *Annalen der Physik und Chemie* 93, no. 12 (1854): 481–506.

———. Über vershiedene für Anwendung bequeme Formen der Hauptgleichungen der mechanishen Wärmetheorie, *Annalen der Physik* 125 (1865): 353–400.

"Climate Change: An Evangelical Call to Action." *Globalist,* February 2, 2007. http://www.theglobalist.com/StoryId.aspx?StoryId=5942.

Cobb, John B. Jr. "Ecology, Science, and Religion: Toward a Postmodern Worldview." In *The Reenchantment of Science: Postmodern Proposals,* edited by David Ray Griffin. Albany: State University of New York Press, 1988.

———. "Process Theology and an Ecological Model." In *Cry of the Environment: Rebuilding the Christian Creation Tradition,* edited by Philip Joranson and Ken Butigan. Santa Fe: Bear, 1984.

Cobb, John B. Jr., and David Ray Griffin. *Process Theology.* Philadelphia: Westminister, 1976.

Cohen, Tom, et al. *Twilight of the Anthropocene Idols.* London: Open Humanities, 2016.

Cox, George. "Alien Species in North America and Hawaii: Impacts on Natural Ecosystems." *Ecology* 81, no. 6 (June 1, 2000): 1756–57.

Cronon, William. "Telling Tales on Canvas: Landscapes of Frontier Change." In Jules Prown et al., *Discovered Lands, Invented Pasts.* New Haven: Yale University Press, 1992.

Crosby, Alfred. *The Columbian Exchange: Biological and Cultural Consequences of 1492.* Westport, CT: Greenwood, 1973.

Crutzen, Paul J. "Geology of Mankind." *Nature* 415, no. 23 (January 3, 2002): 23.

Crutzen, Paul J., Gregor Lax, and Carsten Reinhardt. "Paul Crutzen on the Ozone Hole, Nitrogen Oxides, and the Nobel Prize," December 3, 2012. https://onlinelibrary.wiley.com/doi/full/10.1002/anie.201208700.

Crutzen, Paul J., A. R. Moiser, K. A. Smith, and W. Winiwarter. "$N_2O$ Release from Ago-Biofuel Production Negates Global Warming Reduction by Replacing Fossil Fuels." *Atmos-Chem-Phys, Discuss.* 7 (August 1, 2007): 11191–205.

Crutzen, Paul J., and Eugene F. Stoermer. "The Anthropocene." *IGPB (International Geosphere-Biosphere Programme) Newsletter* 41 (2000): 17.

Curnutt, John L. "A Guide to the Homogenocene, Review of George Cox, 'Alien Species in North America and Hawaii: Impacts on Natural Ecosystems.'" *Ecology* 81, no. 6 (June 1, 2000): 1756–57. doi:10.1890/00129 658(2000)081[1756:AGTTH]2.0.CO%3B2.

Dalakov, Georgi, et al. "The Stepped Reckoner of Gottfried Leibniz." http://history-computer.com/MechanicalCalculators/Pioneers/Lebniz.html.

Davies, Jeremy. *The Birth of the Anthropocene*. Berkeley: University of California Press, 2016.

Davis, Heather, and Etienne Turpin, eds. *Art in the Anthropocene: Encounters among Aesthetics, Politics, Environments and Epistemologies*. London: Open Humanities, 2015.

Davis, Mike. "Living on the Ice Shelf: Humanity's Meltdown," June 26, 2008. htpp://www.tomdispatch.com/post/174949.

Deane-Drummond, Celia, Sigurd Bergmann, and Markus Vogt, eds. *Religion in the Anthropocene*. Eugene, OR: Cascade Books, 2017.

Demos, T. J. "Anthropocene, Capitalocene, Gynocene: The Many Names of Resistance," June 12, 2015. https://www.fotomuseum.ch/en/explore/still-searching/articles/27015_anthropocene_capitalocene_gynocene_the_many_names_of_resistance.

Devall, Bill, and George Sessions, *Deep Ecology*. Salt Lake City: G. M. Smith, 1985.

Diamond, Jared. *Guns, Germs, and Steel: The Fates of Human Societies*. New York: Norton, 1997.

Dickens, Charles. *Dombey and Son* (1848). Edited by Alan Horsman. New York: Oxford University Press, 1966.

———. *Hard Times*. London: Bradbury & Evans, 1854.

Dickinson, Emily. *The Poems of Emily Dickinson*. Boston: Roberts Brothers, 1896.

Dillard, Annie. *Pilgrim at Tinker Creek*. Cutchogue, NY: Buccaneer Books, 1974.

*Earth Charter Initiative*. http://www.earthcharter.org.

Ehlers, Eckart, and Thomas Krafft, eds. *Earth System Science in the Anthropocene: Emerging Issues and Problems*. Berlin: Springer Verlag, 2006.

Eliasson, Olafur. *Your Mobile Expectations: BMW $H_2R$ Project*. Exhibition booklet. Berlin: Studio Olafur Eliasson, 2007.

Ellis, Erle C., Dorian Q. Fuller, Jed O. Kaplan, and Wayne G. Lutters. "Dating the Anthropocene: Towards an Empirical Global History of Human Transformation of the Terrestrial Biosphere." *Elementa: Science of the Anthropocene* 1, no. 18 (December 4, 2018): 1–6.

Elvin, Mark. *The Retreat of the Elephants: An Environmental History of China.* New Haven: Yale University Press, 2006.

Emerson, Ralph Waldo. "The Young American," read before the Mercantile Library Association, Boston, February 7, 1844. www.tamut.edu/academics/mperri/AmSoInHis/The%20Young%20American.doc.

"Entropy and Heat Death." *Encylopaedia Britannica.* https://www.britannica.com/science/thermodynamics/Entropy-and-heat-death.

Enzler, S. M. "History of the Greenhouse Effect and Global Warming." https://www.lenntech.com/greenhouse-effect/global-warming-history.htm.

Erlandson, Jon, and Todd J. Braje, eds. "When Humans Dominated the Earth: Archeological Perspectives on the Anthropocene." *Anthropocene,* January 21, 2013, 1–122. https://www.sciencedirect.com/journal/anthropocene/vol/4.

Eustachewich, Lia. *New York Post,* October 8, 2018. https://www.foxnews.com/science.

Faiola, Anthony. "Pope Francis Presents Trump with a 'Politically Loaded' Gift: His Encyclical on Climate Change." *Washington Post,* May 24, 2017.

Fimrite, Peter. "Dire New Forecast on Global Warming." *San Francisco Chronicle,* November 24, 2018, 1, 10.

Fleming, Donald. "Latent Heat and the Invention of the Watt Engine." *Isis* 43, no. 1 (April 1952): 3–5.

Fox, Robert. *The Caloric Theory of Gases from Lavoisier to Regnault.* Oxford: Clarendon, 1971.

François, Anne-Lise. "Ungiving Time: Reading Lyric by the Light of the Anthropocene." In *Anthropocene: Literary History in Geologic Times,* edited by Tobias Menely and Jesse Oak Taylor. University Park: Pennsylvania State University Press, 2017.

Freese, Barbara. *Coal: A Human History.* New York: Basic Books, 2003.

Frost, Robert. *West-Running Brook.* New York: Henry Holt, 1928.

Gardiner, Stephen. "A Perfect Moral Storm: Climate Change, Intergenerational Ethics and the Problem of Moral Corruption." *Environmental Values* 15 (August 2006): 397–413.

Ghosh, Amitav. *The Circle of Reason.* New York: Houghton Mifflin Mariner Books, 1986.

———. *The Great Derangement: Climate Change and the Unthinkable.* Chicago: University of Chicago Press, 2016.

Glantz, Michael H. "Africans, African-Americans and Climate Impacts: Top-Down vs. Bottom-Up Approach to Capacity Building," July 7, 2006. http://www.fragilecologies.com/jul07_06.html.

Gleick, James. *Chaos Theory: Making a New Science.* New York: Viking, 1987.

"Global Climate Report," April 2018. https://www.ncdc.noaa.gov/sotc/global/201804.

Goldstein, Rebecca Newberger. *Plato at the Googleplex: Why Philosophy Won't Go Away.* New York: Pantheon, 2014.

Goodall, Chris. "Powerful Position: Book Review on *Surviving the Century: Facing Climate Chaos & Other Global Challenges.*" Nature Reports Climate Change. Nature.com 5 (October 12, 2007): 53.

Goodman, Amy. "Native American Activist Winona LaDuke at Standing Rock: It's Time to Move on from Fossil Fuels." *Democracy Now,* September 12, 2016. https://www.democracynow.org/2016/9/12/native_american_activist_winona_laduke_at.

Griffin, David Ray, ed. *The Reenchantment of Science: Postmodern Proposals.* Albany: State University of New York Press, 1988.

———. *Unprecedented: Can Civilization Survive the $CO_2$ Crisis?* Atlanta: Clarity, 2015.

Grim, John, and Mary Evelyn Tucker. *Ecology and Religion.* Washington, DC: Island, 2014.

Grooten, M., and R. E. A. Almond, eds. *Living Planet Report.* Gland, Switzerland: World Wildlife Fund, 2018.

Grusin, Richard, ed. *Anthropocene Feminism.* Center for 21st Century Studies. Minneapolis: University of Minnesota Press, 2017.

Gulliford, Andrew. "Love Stories from Tres Piedras." *Durango Herald,* September 7, 2016. https://www.aldoleopold.org/post/love-stories-tres-piedras/.

Hamilton, Clive. *Defiant Earth: The Fate of Humans in the Anthropocene.* Cambridge: Polity, 2017.

———. *Requiem for a Species*. New York: Routledge, 2015.

Hamilton, Clive, Christophe Bonneuil, and François Gemenne, eds. *The Anthropocene and the Global Environmental Crisis: Rethinking Modernity in a New Epoch*. New York: Routledge, 2015.

Hanania, Jordan, Kailyn Stenhouse, and Jason Donev. "Discovery of the Greenhouse Effect," 2019. http://energyeducation.ca/encyclopedia/Discovery_of_the_greenhouse_effect.

Hanna, Jonathan. "Native Communities and Climate Change: Protecting Tribal Resources as Part of National Climate Policy." Natural Resources Law, Boulder, CO, 2007. http://www.colorado.edu/law/centers/nrlc/publications/ClimateChangeReport-FINAL%20_9.16.07_.pdf.

Haraway, Donna. "Anthropocene, Capitalocene, Plantationocene, and Chthulucene: Making Kin." *Environmental Humanities* 6 (2015): 159–65.

———. *Staying with the Trouble: Making Kin in the Chthulucene*. Durham: Duke University Press, 2016.

———. "Tentacular Thinking: Anthropocene, Capitalocene, Chthulucene." *E-flux Journal*, no. 75 (September 2016).

Harris, Ron. "Law, Finance, and the First Corporations." In *Global Perspectives on the Rule of Law*, edited by James J. Heckman, Robert L. Neilson, and Lee Cabatingan. New York: Routledge, 2009.

———. "The Transplantation of the Legal Discourse on Corporate Personality Theories: From German Codification to British Political Pluralism and American Big Business." *Washington & Lee Law Review* 63 (2006): 1421–78. http://scholarlycommons.law.wlu.edu/wlulr/vol63/iss4/7.

Hawthorne, Nathaniel. *The Celestial Railroad* (1843). http://www.online-literature.com/hawthorne/127/.

———. *The House of the Seven Gables* (1851). New York: Modern Library, 2001.

"Healing Our Planet Earth (HOPE) Conference." https://www.episcopalchurch.org/library/article/faith-based-action-climate-change-urged-seattle-hope-conference.

Hecht, Gabrielle. "The African Anthropocene." *Aeon*, 2018. https://aeon.co/essays/if-we-talk-about-hurting-our-planet-who-exactly-is-the-we.

Hiebert, Erwin. *Historical Roots of the Principle of the Conservation of Energy*. Madison: Wisconsin State Historical Society, 1962.

Horkheimer, Max. *Eclipse of Reason*. New York: Seabury, 1974.

Howarth, Alice. "The Ten Turner Paintings Every Man Needs to See." *GQ Magazine*, October 31, 2015. https://www.gq-magazine.co.uk/article/turner-paintings-top-ten-timothy-spall.

Interfaith Center on Corporate Responsibility. "Priorities of ICCR," 2004. http://www.iccr.org/about/.

IPCC (Intergovernmental Panel on Climate Change). *Climate Change 2001: The Scientific Basis; Contribution of Working Group I to the Third Assessment Report of the Intergovernmental Panel on Climate Change*. Cambridge: Cambridge University Press, 2001.

——. Working Group II Report: "Impacts, Adaptation and Vulnerability"; Working Group III Report, "Mitigation of Climate Change"; Working Group IV Report, "The Physical Science Basis," 2014. http://www.ipcc.ch/.

Jackman, W. T. *The Development of Transportation in Modern England*. 2 vols. Cambridge: Cambridge University Press, 1962.

Jacobson, Mark Z., Mark A. Delucchi, Mary A. Cameron, and Bethany A. Frew. "Low-Cost Solution to the Grid Reliability Problem with 100% Penetration of Intermittent Wind, Water, and Solar for All Purposes." *Proceedings of the National Academy of Sciences* 112, no. 49 (December 8, 2015): 15060–65.

Jantsch, Eric. *The Self-Organizing Universe: Scientific and Human Implications of the Emerging Paradigm of Evolution*. New York: Pergamon, 1980.

Jenkins, Matt. "Carbon Capture." *Nature Conservancy Magazine* (Fall 2018).

"Jet Fuel Consumption by Country." https://www.statista.com/statistics/197690/us-airline-fuel-consumption-since-2004/, p. 3.

Joule, James Prescott. "On Changes of Temperature Produced by the Rarefaction and Condensation of Air." *Philosophical Magazine*, ser. 3, vol. 26, no. 174 (May 1845): 369–83.

Keaten, Jamey. "Melting Ice Opens Route through Arctic." *San Francisco Chronicle*, September 16, 2007, A2.

Kendall-Miller, Heather. Native American Rights Fund News, June 10, 2007. http://narfnews.blogspot.com/2007_06_01_archive.html.

Kingsolver, Barbara. *Flight Behavior*. New York: HarperCollins, 2012.

Klein, Naomi. *This Changes Everything: Capitalism vs. the Climate.* New York: Simon & Schuster, 2014.

Kolbert, Elizabeth. *Field Notes from a Catastrophe: Man, Nature, and Climate Change.* New York: Bloomsbury, 2006.

Kornweibel, Theodore Jr. *Railroads in the African American Experience: A Photographic Journey.* Baltimore: Johns Hopkins University Press, 2010.

Kress, John W., and Jeffrey K. Stine, eds. *Living in the Anthropocene: Earth in the Age of Humans.* Washington, DC: Smithsonian Institution, 2017.

Kuhn, Thomas. *The Copernican Revolution: Planetary Astronomy in the Development of Western Thought.* Cambridge, MA: Harvard University Press, 1957.

———. *The Structure of Scientific Revolutions.* Chicago: University of Chicago Press, 1962.

Kyoto Protocol to the United Nations Framework Convention on Climate Change, 1997. United Nations. *American Journal of International Law* 92, no. 2 (April 1998): 315–31.

Lao Tzu. *The Tao-Teh King.* Translated by C. Spurgeon Medhurst. Wheaton, IL: Theosophical Publishing House, 1972.

Leiserowitz, Anthony. "American Opinions on Global Warming." School of Forestry and Environmental Studies, Yale University, 2007. http://environment.yale.edu/news/5305-american-opinions-on-global-warming/.

———. "American Risk Perceptions: Is Climate Change Dangerous?" *Risk Analysis* 25, no. 6 (2005): 1433–42.

Leopold, Aldo. *A Sand County Almanac.* London: Oxford University Press, 1949.

Lo, Erin. "How Fast Will Jet Fuel Consumption Rise? 2017. https://repository.upenn.edu/cgi/viewcontent.cgi?referer=https://www.google.com/&httpsredir=1&article=1151&context=wharton_research_scholars.

Lomborg, Bjørn. *Cool It: The Skeptical Environmentalist's Guide to Global Warming.* New York: Knopf, 2007.

Lovelock, James. *Gaia: A New Look at Life on Earth.* New York: Oxford University Press, 1979.

Luke, Timothy, ed. "Political Critiques of the Anthropocene." *Telos* 172 (Fall 2015).

Mach, Ernst. *Principles of the Theory of Heat*. Translated by T. J. McCormack. Dordrecht: Reidel, 1986.

Macilenti, Alessandro. *Characterising the Anthropocene: Ecological Degradation in Italian Twenty-First Century Literary Writing*. Berlin: Peter Lang, 2018.

Macintyre, James. "Pope to Make Climate Action a Moral Obligation." *Independent Online*, September 22, 2007. http://news.independent.co.uk/europe/article2987811.ece.

MacKay, Kevin. *Radical Transformation: Oligarchy, Collapse, and the Crisis of Civilization*. Toronto: Between the Lines, 2017.

Magie, William F., trans. and ed. *The Second Law of Thermodynamics: Memoirs by Carnot, Clausius, and Thomson*. New York: Harper, 1899.

——, ed., *A Source Book in Physics*. Cambridge, MA: Harvard University Press, 1963.

Major, Alice. *Welcome to the Anthropocene*. Edmonton: University of Alberta Press, 2018.

Mallet, Whitney. "Naomi Klein's Radical Guide to the Anthropocene." Documentary, 2015. https://newrepublic.com/article/122981/naomi-kleins-radical-guide-anthropocene.

Malm, Andreas. "The Anthropocene Myth." *Jacobin Magazine*, March 30, 2015.

——. *Fossil Capital: The Rise of Steam Power and the Roots of Global Warming*. London: Verso, 2016.

Mann, Charles. "The Dawn of the Homogenocene." *Orion Magazine*, 2011. https://orionmagazine.org/article/the-dawn-of-the-homogenocene/.

——. "Living in the Homogenocene: The First 500 Years." The Long Now Foundation, April 23, 2012. http://longnow.org/seminars/02012/apr/23/living-homogenocene-first-500-years/.

——. *1493: Uncovering the New World Columbus Created*. New York: Knopf, 2011.

Martyris, Nina. "Barbara Kingsolver, Barack Obama, and the Monarch Butterfly." *New Yorker*, April 10, 2015.

Marx, Leo. *The Machine in the Garden: Technology and the Pastoral Ideal in America.* New York: Oxford University Press, 1967.

Maslin, Mark. *Global Warming: A Very Short Introduction.* New York: Oxford University Press, 2004.

Maxwell, James C. *Theory of Heat.* New York: Dover, 1871.

McDaniel, Jay. "Christian Spirituality as Openness to Fellow Creatures." *Environmental Ethics* 8, no. 4 (Spring 1986): 33–46.

———. *Of God and Pelicans: A Theology of Reverence for Life.* Louisville, KY: Westminster John Knox, 1989.

———. "Physical Matter as Creative and Sentient." *Environmental Ethics* 5, no. 4 (Winter 1983): 291–317.

McDonald, Norris. "Global Climate Change and the African-American Community (Part I)." African American Environmentalist Association, 2007. http://www.aaenvironment.com/GlobalWarming1.htm.

McEvoy, Paul. *Classical Theory.* San Francisco: Microanalytix, 2002.

McKibben, Bill. "Can Anyone Stop It?" *New York Review of Books,* October 11, 2007.

———. *The End of Nature.* New York: Random House, 1989.

McNeill, John R. "Nature Preservation and Political Power in the Anthropocene." In *After Preservation: Saving American Nature in the Age of Humans,* edited by Ben Minteer and Stephen J. Pyne. Chicago: University of Chicago Press.

McNeill, John R., and Peter Engelke. *The Great Acceleration: An Environmental History of the Anthropocene since 1945.* Cambridge, MA: Harvard University Press, 2014.

McPhee, John. "Coal Train—A Reporter at Large," part 1. *New Yorker,* October 3, 2005; part 2, October 10, 2005.

Mendoza, Eric, ed. *Reflections on the Motive Power of Fire by Sadi Carnot and Other Papers on the Second Law of Thermodynamics by E. Clapeyron and R. Clausius.* New York: Dover, 1960.

Mentz, Steven. "Anthropocene v. Homogenocene," January 25, 2013. http://stevementz.com/anthropocene-v-homogenocene/.

Merchant, Carolyn. *Autonomous Nature: Problems of Prediction and Control from Ancient Times to the Scientific Revolution*. New York: Routledge, 2016.

———. *The Death of Nature: Women, Ecology, and the Scientific Revolution*. 3rd ed. San Francisco: HarperCollins, 2020.

———. *Earthcare: Women and the Environment*. New York: Routledge, 1996.

———. *Ecological Revolutions: Nature, Gender and Science in New England*. 2nd ed. Chapel Hill: University of North Carolina Press, 2010.

———. https://ourenvironment.berkeley.edu/people/carolyn-merchant.

———, ed. *Major Problems in American Environmental History: Documents and Essays*. 3rd ed. Boston: Wadsworth Cengage, 2012.

———. "Partnership Ethics: Business and the Environment." In *Environmental Challenges to Business*, edited by Patricia Werhane. 1997 Ruffin Lectures, University of Virginia Darden School of Business. Bowling Green, OH: Society for Business Ethics, 2000.

———. "Partnership with Nature." In "Eco-Revelatory Design: Nature Constructed/Nature Revealed." Special issue, *Landscape Journal* (1998): 69–71.

———. *Radical Ecology: The Search for a Livable World*. 2nd ed. New York: Routledge, 2005.

———. *Reinventing Eden*. 2nd ed. New York: Routledge, 2013.

———. "Restoration and Reunion with Nature." *Restoration and Management Notes* 4 (Winter 1986): 68–70.

———. *Science and Nature: Past, Present, and Future*. New York: Routledge, 2018.

Miller, Brandon, and Jay Croft. "'Life-or-Death' Warning: Major Study Says World Has Just 11 Years to Avoid Climate Change Catastrophe." *CNN*, October 8, 2018. https://amp.cnn.com/cnn/2018/10/07/world/climate-change-new-ipcc-report-wxc/index.html.

Mirzoeff, Nicholas. "Visualizing the Anthropocene." *Public Culture* 26, 2 (2014): 213–32.

Moore, Jason, ed. *Anthropocene or Capitalocene? Nature, History, and the Crisis of Capitalism*. Oakland, CA: PM Press, Kairos Books, 2016.

———. *Capitalism in the Web of Life: Ecology and the Accumulation of Capital*. London: Verso, 2015.

———. "The Capitalocene: On the Nature and Origins of Our Ecological Crisis," pt. 1, "The Capitalocene: Abstract Social Nature and the Limits to Capital." *Journal of Peasant Studies* 44, no. 3 (2017): 594–630.

———. "The Capitalocene," pt. 2, "Accumulation by Appropriation and the Centrality of Unpaid Work/Energy." *Journal of Peasant Studies*. doi:10.1080/03066150.2016.1272587.

———. *Ecology in the Web of Life: Ecology and the Accumulation of Capital*. London: Verso, 2015.

Morello-Frosch, Rachel, and Bill M. Jesdale. "Separate and Unequal: Residential Segregation and Estimated Cancer Risks Associated with Ambient Air Toxics in U.S. Metropolitan Areas." *Environmental Health Perspectives* 114, no. 3 (March 2006): 386–93.

Morrison, Alex. "Envisioning Change: Combating Climate Change with Art." *PFSK*, July 26, 2007. http://www.psfk.com/2007/07/envisioning-change-combating-climate-change-with-art.html.

Mullan, John. "My Favorite Dickens: *Dombey and Son*." *Guardian*, September 23, 2011. https://www.theguardian.com/books/2011/sep/23/charles-dickens-favourite-dombey-son.

———. "Railways in Victorian Fiction." Discovering Literature: Romantics and Victorians, British Library, May 15, 2014. https://www.bl.uk/romantics-and-victorians/articles/railways-in-victorian-fiction#.

Naess, Arne. "The Shallow and the Deep, Long-Range Ecology Movement." *Inquiry* 16 (1973): 95–100.

Nahm, Milton C., ed. *Selections from Early Greek Philosophy*. 3rd ed. New York: Appleton-Century-Crofts, 1947.

Nebeky, Tobias, and Jesse Oak Taylor, eds. *Anthropocene Reading: Literary History in Geologic Times*. University Park: Pennsylvania University Press, 2017.

Needham, Joseph. *Science and Civilization in China*. Cambridge: Cambridge University Press, 1956.

Newburgh, Ronald. "Carnot to Clausius: Caloric to Entropy." *European Journal of Physics* 30 (2009): 713–28.

Newton, Isaac. *Mathematical Principles of Natural Philosophy*. Translated by A. Motte and F. Cajori. Berkeley: University of California Press, 1960.

Nordhaus, Ted, and Michael Shellenberger. *Break Through: From the Death of Environmentalism to the Politics of Possibility.* Boston: Houghton Mifflin, 2007.

"Ocean Acidification." *National Geographic,* April 17, 2017.

Paavola, Jouni, and W. Neil Adger. *Fairness in Adaptation to Climate Change.* Cambridge, MA: MIT Press, 2006.

Palmer, Donald. *Looking at Philosophy: The Unbearable Heaviness of Philosophy Made Lighter.* Mountain View, CA: Mayfield, 1988.

Pálsson, Gísli, Sverker Sörlin, Brownislaw Szerzynski, et al. "The Anthropocene: Integrating the Social Sciences and Humanities." *Environmental Science and Policy* 28 (2013): 4–13.

Parker, Alan, et al. "Climate Change and Pacific Rim Indigenous Nations." Northwest Indian Applied Research Institute. Olympia, WA: Evergreen State College, October 2006. http://academic.evergreen.edu/g/grossmaz/IndigClimate2.pdf.

Patel, Prachi. "Airplanes Flying on Biofuels Emit Fewer Climate-Warming Particles." Anthropocenemagazine.org, March 16, 2016. http://www.anthropocenemagazine.org/2017/03/airplanes-flying-on-bio-jetfuel-emit-fewer-climate-warming-particles/.

Pereira Savi, Melina. "The Anthropocene (and) (in) the Humanities," *Revista estudos feministas* 25, no. 2 (May–August 2017.)

Pope Francis. "Laudato Si: On Care for Our Common Home." May 24, 2015. http://m.vatican.va/content/francescomobile/en/encyclicals/documents/papa-francesco_20150524_enciclica-laudato-si.html.

Prigogine, Ilya. "Time, Structure, and Fluctuations." Nobel lecture, December 8, 1977. https://www.nobelprize.org/nobel_prizes/chemistry/laureates/1977/prigogine-lecture.pdf.

Prigogine, Ilya, and Isabelle Stengers. *Order out of Chaos: Man's New Dialogue with Nature.* New York: Bantam, 1984.

Purdy, Jedediah. *After Nature: A Politics for the Anthropocene.* Cambridge, MA: Harvard University Press, 2015.

Raftery, Adrian E., et al. "Less Than 2°C Warming by 2100 Unlikely." *Nature Climate Change* 7 (September 2017): 637–41.

# BIBLIOGRAPHY

Rankine, William. *A Manual of the Steam Engine and Other Prime Movers.* London: R. Griffin, 1859.

———. "On the General Law of the Transformation of Energy." *London, Edinburgh, and Dublin Philosophical Magazine and Journal of Science,* ser. 4, vol. 5 (1853): 106–17.

———. "On the Mechanical Action of Heat, Especially in Gases and Vapours." *Transactions of the Royal Society of Edinburgh* 20 (February 4, 1850): 147–64.

Rawls, John. *A Theory of Justice.* Oxford: Oxford University Press, 1971.

Rayner, Steve, and Elizabeth L. Malone, eds. *The Societal Framework (Human Choice and Climate Change).* Vol. 1. Columbus: Batell, 1998.

Ritvo, Harriet. "Fighting for Thirlmere: The Roots of Environmentalism." *Science* 300 (June 6, 2003): 1510–11.

Robin, Libby, Sverker Sörlin, and Paul Warde, eds. *The Future of Nature: Documents of Global Change.* New Haven: Yale University Press, 2013.

Rockeymore, Maya. *African-Americans and Climate Change: Unequal Burdens and Ethical Dilemmas.* Washington, DC: Congressional Black Caucus Foundation, 2007.

Ruddiman, William F. "The Anthropogenic Greenhouse Era Began Thousands of Years Ago." *Climatic Change* 61, no. 3 (2003): 261–93.

———. "Debate over the Early Anthropogenic Hypothesis." *RealClimate,* December 2005.

———. *Earth Transformed.* New York: W. H. Freeman, 2013.

———. "How Did Humans First Alter Global Climate?" *Scientific American,* March 2005.

Ruether, Rosemary Radford. *Integrating Ecofeminism, Globalization and World Religions.* Lanham, MD: Roman & Littlefield, 2005.

Runes, Dagobert D. *Pictorial History of Philosophy.* New York: Philosophical Library, 1959.

Russell, Dick. "Environmental Racism: Minority Communities and Their Battle against Toxics." *Amicus Journal* 11, no. 2 (Spring 1989): 22–32.

Said, Carolyn. "Uber Is on the Road to Becoming the Amazon of Transportation." *San Francisco Chronicle,* September 7, 2018, C-1, C-3.

Samways, Michael. "Translocating Fauna to Foreign Lands: Here Comes the Homogenocene." *Journal of Insect Conservation* 3 (1999): 65–66.

Schivelbusch, Wolfgang. *The Railway Journey: The Industrialization of Time and Space in the Nineteenth Century*. 3rd ed. Berkeley: University of California Press, 2014.

Schwartz, Robert. Mt. Holyoke College, "The Industrial Revolution and the Railroad System," "Opposing Voices," https://www.mtholyoke.edu/courses/rschwart/ind_rev/voices/wordsworth.html.

Sellers, Charles. *The Market Revolution: Jacksonian America, 1815–1846*. New York: Oxford University Press, 1991.

Shedd, John C. "A Mechanical Model of the Carnot Engine." *Physical Review* 8, no. 3 (January 1899): 174–80.

Shepard, John. "Can't We Just Remove Carbon Dioxide from the Air to Fix Climate Change?" August 3, 2015. http://theconversation.com/cant-we-just-remove-carbon-dioxide-from-the-air-to-fix-climate-change-not-yet-45621.

Sherma, Rita D., and Arvind Sharma, eds. *Hermeneutics and Hindu Thought: Toward a Fusion of Horizons*. New York: Springer, 2008.

Simon, Kathy. "Railroad Paintings and Art." https://www.pinterest.com/xbowler/railroad-paintings-and-art/.

Singer, Peter. "Ethics and Climate Change: Commentary." *Environmental Values* 15, no. 3 (2006): 415–22.

Smith, Crosbie, and Norton Wise. *Energy and Empire: A Biographical Study of Lord Kelvin*. New York: Cambridge University Press, 1989.

Snyder, Gary. *Riprap and Cold Mountain Poems*. Berkeley: Counterpoint, 2009.

Solnick, Sam. *Poetry and the Anthropocene: Ecology, Biology, and Technology in Contemporary British and Irish Poetry*. New York: Routledge, 2016.

Soni, Jimmy, and Rob Goodman. *How Claude Shannon Invented the Information Age*. New York: Simon & Schuster, 2017.

Sörlin, Sverker. "Environmental Turn in the Human Sciences" and "The Anthropocene: What Is It?" *The Institute Letter* (Institute for Advanced Study, Princeton, NJ) (Summer 2014): 1, 12–13.

Spangenberg, Joachim. "China in the Anthropocene: Culprit, Victim or Last Best Hope for a Global Ecological Civilization?" *BioRisk* 9 (2014): 1–37.

"Steam Engine." Wikipedia. http://www.deutsches-museum.de/en/information/young-people/inventors-trail/drivetrains/steam-engine/.

Steffen, Will, Paul Crutzen, and John McNeill. "The Anthropocene: Are Humans Now Overwhelming the Great Forces of Nature?" *Ambio* 6, no. 8 (December 2007): 614–71.

Steffen, Will, et al. *Global Change and the Earth System: A Planet under Pressure.* New York: Springer, 2004.

Stevens, Lara, Peta Tait, and Denise Varney, eds. *Feminist Ecologies: Changing Environments in the Anthropocene.* London: Palgrave MacMillan, 2018.

Stewart, Jill. "Home May Rise on Incinerator Site." *Los Angeles Times,* May 30, 1990. http://articles.latimes.com/1990-05-30/news/mn-277_1_los-angeles.

Taylor, Bron, ed. *Encyclopedia of Religion and Nature.* New York: Bloomsbury, 2005.

Thomas, Inigo. "The Chase." *London Review of Books,* October 20, 2016, 15–18. https://www.lrb.co.uk/v38/n20/inigo-thomas/the-chase.

Thomson, William (Lord Kelvin). *Mathematical and Physical Papers.* Cambridge: Cambridge University Press, 1882.

———. "On the Dynamical Theory of Heat with Numerical Results Deduced from Mr. Joule's Equivalent of a Thermal Unit and M. Regnault's *Observations on Steam.*" *Transactions of the Royal Society of Edinburgh* (March 1851): 8–21, 105–17, 168–76.

———. "On the Universal Tendency in Nature to the Dissipation of Mechanical Energy." *Proceedings of the Royal Society of Edinburgh* 4 (April 19, 1852): 304–6.

Thoreau, Henry David. *Walden; or, Life in the Woods.* Boston: Ticknor & Fields, 1854.

Thorne, Tony. *The Singularity Is Coming . . . ! The Artificial Intelligence Explosion.* N.p.: CreateSpace, 2015.

Tory, Sarah. "Religious Communities Are Taking on Climate Change." *High Country News,* September 18, 2017. http://www.hcn.org/issues/49.16/activism-why-religious-communities-are-taking-on-climate-change.

Totman, Conrad. *Japan: An Environmental History*. London: I. B. Tauris, 2014.

Trexler, Adam. *Anthropocene Fictions: The Novel in a Time of Climate Change*. Charlottesville: University of Virginia Press, 2015.

Tsing, Anna, Heather Swanson, Elaine Gan, and Nils Bubandt, eds. *Arts of Living on a Damaged Planet: Ghosts and Monsters of the Anthropocene*. Minneapolis: University of Minnesota Press, 2017.

Tucker, Mary Evelyn. *The Emerging Alliance of Religion and Ecology*. Salt Lake City, UT: University of Utah Press, 2014.

Tucker, Mary Evelyn, and John Grim, eds. *Religions of the World and Ecology*. 9 vols. Cambridge, MA: Harvard University Press, 1997–2004.

Tyndall Centre for Climate Change Research. "Justice and Adaptation to Climate Change." Working Paper 23, October 2002. http://www.tyndall.ac.uk/publications/working_papers/wp23.pdf.

Union of Concerned Scientists. "Capping Global Warming Emissions." California Climate Choices, a Fact Sheet, 2006. http://www.law.stanford.edu/program/centers/enrlp/pdf/AB-32-fact-sheet.pdf.

United Nations Environment Program. "Envisioning Change: Melting Ice/Hot Topic." Art for the Environment Project for World Environment Day. Oslo, Norway, 2007. https://www.culturenet.hr/UserDocsImages/attachmenti/15522.pdf.

United States Global Change Research Program. "United States National Assessment of the Potential Consequences of Climate Variability and Change Region: Native Peoples/Native Homelands," May 25, 2005. http://www.usgcrp.gov/usgcrp/nacc/npnh-sw.htm.

U.S. Commission on Civil Rights. "Not in My Backyard, Executive Order 12,898 and Title VI as Tools for Achieving Environmental Justice," October 2003. http://www.usccr.gov/pubs/envjust/ej0104.pdf.

U.S. Environmental Protection Agency. Climate Change Science. "Future of Climate Change," 2017. https://19january2017snapshot.epa.gov/climate-change-science/future-climate-change_.html.

Vidal, John, and Tom Kington. "Pope Puts Focus on Climate Change and the Environment." *The Age Newspaper* (Melbourne), April 28, 2007.

# BIBLIOGRAPHY

Vince, Gaia. *Adventures in the Anthropocene: A Journey to the Heart of the Planet We Made.* Minneapolis: Milkweed, 2014.

Viveiros de Castro, Eduardo. "Exchanging Perspectives." *Common Knowledge* 10, no. 3 (Fall 2004): 463–84.

Voosen, Paul. "Scientists Drive Golden Spike toward Anthropocene's Base." *Greenwire,* September 17, 2012. https://www.eenews.net/stories/1059970036.

Waldman, Anne. "Anthropocene Blues," 2017. Originally published in *Poem-a-Day* on February 2, 2016, by the Academy of American Poets. https://www.poets.org/poetsorg/poem/anthropocene-blues.

Waldrop, M. Mitchell. *Complexity: The Emerging Science at the Edge of Order and Chaos.* New York: Simon & Schuster, 1992.

Wark, McKenzie. *Molecular Red: Theory for the Anthropocene.* London: Verso, 2015.

Waters, Colin N., et al. "Can Nuclear Weapons Fallout Mark the Beginning of the Anthropocene Epoch?" *Bulletin of the Atomic Scientists* 7, no. 3 (2015): 46–57.

Watts, Jonathan. "'For Us the Land Is Sacred:' On the Road with the Defenders of the World's Forests." *Guardian,* November 4, 2017. https://www.theguardian.com/environment/2017/nov/04/bonn-climate-conference-on-the-road-with-defenders-of-the-forest.

Weart, Spencer. *The Discovery of Global Warming.* Cambridge, MA: Harvard University Press, 2003.

———. *The Discovery of Global Warming: A History.* https://history.aip.org/climate/summary.htm.

Wells, Jennifer. *Complexity and Sustainability.* New York: Routledge, 2013.

Wells, Jennifer, and Carolyn Merchant. "Melting Ice: Climate Change and the Humanities." *Confluence* 14, no. 2 (Spring 2009): 13–27.

"What Exactly Is the Heat Death of the Universe and Where Can I Find out More?" http://www.physlink.com/education/askexperts/ae181.cfm.

White, Lynn Jr. "The Historical Roots of Our Ecologic Crisis." *Science* 155, no. 3767 (March 10, 1967): 1203–7.

Whitehead, Alfred North. *Process and Reality.* Edited by David Ray Griffin and Donald W. Sherburne. New York: Free Press, 1978.

Whitman, Walt. "To a Locomotive in Winter." https://www.poets.org/poets org/poem/locomotive-winter.

Wilcox, Shari. "Resisting the Plantationocene: The Case of Postcolonial and Post-slavery Banana Plantations in the French Caribbean." University of Wisconsin, Madison, February 27, 2018. https://sts.wisc.edu/event/resisting-the-plantationocene-the-case-of-postcolonial-and-post-slavery-banana-plantations-in-the-french-caribbean/.

Wilson, Steven S. "Sadi Carnot." *Scientific American* 245, no. 2 (August 1981): 131–45.

Wordsworth, William. *The Collected Poems of William Wordsworth*. Edited by Antonia Till. Wordsworth Poetry Library. Hertfordshire, UK: Wordsworth Editions, 1994.

Worster, Donald. *Nature's Economy: A History of Ecological Ideas*. New York: Cambridge University Press, 1994.

Worthy, Kenneth, Elizabeth Allison, and Whitney A. Bauman, eds. *After the Death of Nature: Carolyn Merchant and the Future of Human-Nature Relations*. New York: Routledge, 2018.

Wrigley, Tony. *Continuity, Chance and Change: The Character of the Industrial Revolution in England*. Cambridge: Cambridge University Press 1990.

———. *Energy and the English Industrial Revolution*. Cambridge: Cambridge University Press, 2010.

Yaqoob, M. Mateen. "Introduction to Computers, History and Applications." https://slideplayer.com/slide/8887437/.

Yohe, Gary. "An Issue of Equity." Book review of *Fairness in Adaptation to Climate Change*, by W. Neil Adger, Jouni Paavola, Saleemul Huq, and M. J. Mace. *Nature Reports Climate Change* 5 (October 2007). doi:10.1038/climate.2007.51.

Zalasiewicz, Jan, Mark Williams, Will Steffen, and Paul Crutzen. "The New World of the Anthropocene." *Environmental Science and Technology*, 44 (2010): 2228–31.

# Illustration Credits

### Introduction

Figure I.1. Paul Crutzen: Getty Images.
Figure I.2. Eugene Stoermer: Photograph provided by Russell G. Kreis, Jr.
Figure I.3. The Holocene: International Geosphere-Biosphere Programme, http://www.igbp.net/globalchange/anthropocene.4.1b8ae20512db692f 2a680009238.html. Public domain.
Figure I.4. Global temperature change, 1880–2010: "Global Warming and the Climate," http://www.global-warming-and-the-climate.com/green house-warming-argument.html. Public domain.
Figure I.5. The Human Footprint: from Jessica Stites, "The Dawning of the Age of the Anthropocene," *In These Times* (Apr. 14, 2014): 1, chart designed by Rachel K. Dooley, from "The Dawning of the Age of the Anthropocene," *In These Times*, copyright 2014, used by permission.
Figure I.6. Environmental Protection Agency's projected atmospheric greenhouse gas concentrations, 2000–2100: https://19january2017 snapshot.epa.gov/climate-change-science/future-climate-change _.html. Public domain.
Figure I.7. Svante Arrhenius: The Nobel Foundation, http://nobelprize .org/nobel_prizes/chemistry/laureates/1903/arrhenius-bio.html, Wikimedia Commons, http://commons.wikimedia.org/wiki/File:Svante _Arrhenius.jpg#/media/File:Svante_Arrhenius.jpg. Public domain.
Figure I.8. IGBP, the Great Acceleration: courtesy of Will Steffen.
Figure I.9. Donna Haraway: Courtesy of Donna Haraway.

ILLUSTRATION CREDITS

Figure I.10. Dipesh Chakrabarty: Courtesy of Dipesh Chakrabarty and Alan Thomas.

Figure I.11. Naomi Klein: photograph by Jay L. Clendenin; courtesy of *Los Angeles Times.*

Figure I.12. Ian Angus: Courtesy of Ian Angus.

Figure I.13. Eduardo Viveiros de Castro: Courtesy of Eduardo Viveiros de Castro.

Figure I.14. Jason W. Moore: Courtesy of Jason W. Moore.

## Chapter 1

Figure 1.1. Newcomen engine: Courtesy of Joseph Siry. Public domain.

Figure 1.2. James Watt: Public domain.

Figure 1.3. James Watt steam engine: Deutsches Museum, Munich.

Figure 1.4. Sadi Carnot: Public domain.

Figure 1.5. Benoît Paul Émile Clapeyron: Public domain.

Figure 1.6. Rudolf Clausius: Public domain.

Figure 1.7. William Thomson (Lord Kelvin): Chris Hellier/ Alamy Stock Photo.

Figure 1.8. William Rankine: Public domain.

Figure 1.9. Ludwig Boltzmann: Public domain.

Figure 1.10. Boltzmann's equation for entropy: https://creativecommons.org/licenses/by-sa/3.0/.

## Chapter 2

Figure 2.1. Steam engine crossing the landscape: Greg Kelton / Alamy Stock Photo.

Figure 2.2. Stationary steam engine: Creative Commons Attribution Share-alike license 2.0.

Figure 2.3. Joseph Turner: Tate Gallery, London. Public domain.

Figure 2.4. Joseph Turner, *Fighting Temeraire Being Tugged to Her Last Berth,* 1838: National Gallery, London. Public domain.

Figure 2.5. Claude Monet, *Arrival of the Normandy Train,* 1877: Art Institute of Chicago, Mr. and Mrs. Martin A. Ryerson Collection, ref.

## ILLUSTRATION CREDITS

no. 1933.1158, https://www.artic.edu/artworks/16571/arrival-of-the-normandy-train-gare-saint-lazare, https://www.artic.edu/image-licensing. Public domain.

Figure 2.6. A train barrels down the tracks: Public domain, United States.

Figure 2.7. "The 1833 Steamboat New England," 1919: Essex Institute, The Essex Institute Historical Collections, Peabody Essex Museum, 1859–1993 (Salem, MA: Essex Institute Press), vol. 55, p. 128. Public domain.

Figure 2.8. South Boston Iron Company, engraving 1884: PRISMA ARCHIVO / Alamy Stock Photo.

Figure 2.9. Andrew Melrose, *Westward the Star of Empire Takes Its Way*, 1867: Museum of the American West, Los Angeles, Jane Cazneau Archive, https://janecazneau.omeka.net/items/show/16. Public domain.

Figure 2.10. John Gast, *American Progress*, 1872: Autry Museum of the American West, Los Angeles. Public domain.

Figure 2.11. John Kane, *The Monongahela River Valley, Pennsylvania*, 1931. © The Metropolitan Museum of Art. Image Source: Art Resource, NY.

Figure 2.12. Trackwomen at the Baltimore & Ohio Railroad Company, 1943: National Archives, Research.archives.gov/description/522888. Public domain.

Figure 2.13. *African American Railway Workers:* photograph by Cicero C. Simmons, courtesy Theodore Kornweibel Collection, California State Railroad Museum Library, Sacramento.

Figure 2.14. Female engineer, Llangollen Railway, Wales: 2ebill / Alamy Stock Photo.

Figure 2.15. Olafur Eliasson, *Your Mobile Expectations: BMW $H_2R$ Project*, 2007, On behalf of Olafur Eliasson; © Olafur Eliasson, used by permission.

## Chapter 3

Figure 3.1. William Wordsworth: IanDagnall Computing / Alamy Stock Photo.

Figure 3.2. The *William Wordsworth:* http://www.davidheyscollection.com/userimages/00-0-a-rk-blencowe-70030-folkestone.jpg, copyright © Rod Blencowe (r.blencowe@ntlworld.com), used by permission.

Figure 3.3. Charles Dickens: © Victoria and Albert Museum, London.

Figure 3.4. Nathaniel Hawthorne: Charles Osgood (American, 1809–1890). Portrait of Nathaniel Hawthorne, 1840. Oil on canvas. Salem, Massachusetts, United States. 29½ × 24½ inches (74.93 × 62.23 cm). Peabody Essex Museum, Gift of Professor Richard C. Manning, 1933. 121459. Courtesy of Peabody Essex Museum. Photo by Mark Sexton.

Figure 3.5. Ralph Waldo Emerson: Public domain.

Figure 3.6. Henry David Thoreau: Public domain.

Figure 3.7. Walden Train Station, or View of the Pavilion at Walden Pond: unknown American artist, undated, courtesy Concord Free Public Library, William Munroe Special Collections.

Figure 3.8. Mark Twain: *Encylopaedia Britannica,* https://www.britannica.com/biography/Mark-Twain/media/610829/138635; Prints and Photographs Division/Library of Congress, Washington, DC. (neg. no. LC-USZ62–5513). Public domain.

Figure 3.9. Walt Whitman: Public domain.

Figure 3.10. Emily Dickinson: IanDagnall Computing / Alamy Stock Photo.

Figure 3.11. Annie Dillard: Phyllis Rose. Reproduced courtesy of Russell & Volkening as agents for the author.

## Chapter 4

Figure 4.1. John Grim and Mary Evelyn Tucker: Courtesy of Mary Evelyn Tucker.

Figure 4.2. Pope Benedict XVI. From website of President of Republic of Poland, free documentation license.

## Chapter 5

Figure 5.1. Google Headquarters parking lot, "Electric Vehicle Only": Photograph by Carolyn Merchant.

ILLUSTRATION CREDITS

Figure 5.2. Google Headquarters parking lot, "Expectant Mother Parking": Photograph by Carolyn Merchant.

Figure 5.3. Google Headquarters: Photograph by Carolyn Merchant.

Figure 5.4. Carolyn Merchant's visit to Google Headquarters carrying *Plato at the Googleplex:* Photograph by Carolyn Merchant.

Figure 5.5. Socrates descending from the clouds in a basket: Public domain.

Figure 5.6. Plato and Aristotle: Public domain, United States.

Figure 5.7. Heraclitus of Ephesus: Philosophical Library, Inc., used by permission.

Figure 5.8. Parmenides of Elea: Philosophical Library, Inc., used by permission.

Figure 5.9. Empedocles of Akragas: Philosophical Library, Inc., used by permission.

Figure 5.10. Democritus of Abdera: Philosophical Library, Inc., used by permission.

Figure 5.11. Pythagoras of Samos: Philosophical Library, Inc., used by permission.

Figure 5.12. Isaac Newton (1642–1727): By kind permission of the Trustees of the Portsmouth Estates, UK, Parish of Farleigh.

Figure 5.13. Gottfried Wilhelm Leibniz: Courtesy of akg-images.

Figure 5.14. Albert Einstein: Philosophical Library, Inc., used by permission.

Figure 5.15. Edward Lorenz: Fair use.

Figure 5.16. Ilya Prigogine: Public domain.

## Chapter 6

Figure 6.1. John Locke: Philosophical Library, Inc., used by permission.

Figure 6.2. Jeremy Bentham: Welcome Collection, https://wellcomecollection.org/works/qcugdr6w?query=Jeremy+Bentham. Public domain.

Figure 6.3. John Stuart Mill: Philosophical Library, Inc., used by permission.

ILLUSTRATION CREDITS

Figure 6.4. Aldo Leopold: Courtesy of the Aldo Leopold Foundation, www.aldoleopold.org.

Figure 6.5. Peter Singer: Photograph by Alletta Vaandering, used by permission of Peter Singer and Alletta Vaandering.

Figure 6.6. Stephen M. Gardiner: University of Washington Photography, used by permission.

Figure 6.7. Warren County protest: image by Jerome Friar, 1982, in the Jerome Friar Photographic Collection and Related Materials (P0090), North Carolina Collection, University of North Carolina Library at Chapel Hill.

Figure 6.8. Protest over proposed sewage plant: David Vita, used by permission.

Figure 6.9. Robert Bullard: Courtesy of Robert Bullard.

## Epilogue

Figure E.1. Mark Jacobson: Courtesy of Mark Jacobson.

Figure E2. Christopher Clack: Photograph from Cooperative Institute for Research in Environmental Sciences (CIRES), University of Colorado, Boulder, CIRES NOAA. Public domain.

Figure E.3. Ecological Revolutions diagram: Carolyn Merchant, "The Theoretical Structure of Ecological Revolutions," *Environmental Review* 11, no. 4 (Winter 1987): 268.

# Index

Page numbers in *italics* refer to illustrations.

acid rain, 151
African Americans, 60, *61*, 135, 136, 137, 139, 140, 142
agriculture, 3, 13, 14, 44, 141; greenhouse effect and, 150; large-scale, 15; sustainable, 145
airplanes, 45, 85, 149
air pollution, x, 47, 99, 101, 129
Alaska Natives, 141–42
Alembert, Jean Le Rond d', 27
*American Progress* (Gast), 56–57
Anaxagoras, 116
Anaximenes, 114–15
Androcene, 60, 87
Angus, Ian, 19
*Annals of Physics*, 36
Antarctic, xii, 11, 25
Anthropocene, 17–18, 45; advent of, 26–27; Capitalocene vs., 21–24; climate justice linked to, 143; debated meaning of, 13–14; environmental ethics and, 127–35; environmental humanities in, x–xi, xii; fossil fuels linked to, x, 14, 24–25, 27, 113–14; fundamental dilemma of, 112; initial conception of, x, 1–2, 14, 26; labor in, 59–61; in non-Western world, 25; philosophical questions in, 114–19; prospects for, xii, 144–56; steam engine linked to, x, 2, 26, 152
"Anthropocene Blues" (Waldman), 86
*Anthropocene Fictions* (Trexler), 86
*Anthropocene or Capitalocene?* (Moore), 21–22
appropriate technology, 153
Arctic, xii, 9, 11, 25, 141
Arctic Native peoples, 141
Aristophanes, 109
Aristotle, 112, *113*, 119
Arrhenius, Svante, 6, *7*
*The Arrival of a Train at La Ciotat Station* (Lumière brothers), 52
*Arrival of the Normandy Train* (Monet), 51–52
art, 46–65
Association for the Study of Literature and the Environment (ASLE), 86–87
atomist philosophy, 116, 119, 121
Australia, 10, 148

# INDEX

automobiles, 12–13, 45, 84–85; electric and self-driving, 62, 85
aviation, 45, 85, 149

Babbage, Charles, 120
Baha'i faith, 92
barley, 3
Bartholomew, Greg, 79
Bateson, Gregory, 154
beans, 3, 14
Benedict XVI, Pope, 95, *96*
Bennett, Jane, 87–88
Bentham, Jeremy, 129
Berndt, Brooks, 104–5
biodiversity, 15, 130, 146
biofuels, 85–86, 149
bioregionalism, 130, 145, 154
Black, Joseph, 27
black holes, ix
*Black Metropolis* (Bullard), 138–39
Blackstone Canal, 53
Bohm, David, 155
Bohr, Niels, 121
Boltzmann, Ludwig, 42, *43*
Bookchin, Murray, 154
Botkin, Daniel, 124
Boulton, Matthew, 30
Bouzid, Ahmed, 105
Boyle, Robert, 91, 126
Bradford, William, 56
Brondmo, Hans Peter, 107
Brown, Jerry, 11
Brown, Joan, 104
Bruntland, Gro Harlem, 145
Buckland, David, 64
Buddhism, 92, 98
Bullard, Robert, 138–39
*Bulletin of the Atomic Scientists*, 14
"butterfly effect," 122

calculus, 119, 123
California, 10
Callicott, J. Baird, 130
caloric theory, 34, 35, 39
canals, 52–53
Cape Farewell Project, 63–64
capitalism, 13, 18, 19–24, 26, 44, 146, 150, 156
Capitalocene, 16, 19–24, 150
carbon, 132; capture of, 11, 12, 149–50
carbon dioxide ($CO_2$) emissions: increasing levels of, 3, 4, 5, 27, 45, 150; misconceptions surrounding, 7; policies and methods to reduce, 6, 11–12; in rich vs. poor countries, 20
Caring for Creation, 105
Carnot, Hippolyte, 31
Carnot, Lazare, 31
Carnot, Sadi, 31–32, 35, 36, 42
Carnot cycle, 33, 35
Castillo, Aurora, 136
Catholicism, 95–96
*The Celestial Railroad* (Hawthorne), 72–74
cement industry, 149
Center for Earth Ethics, 95
Center for the Study of World Religions, 93
Central Pacific Railroad, 55
CFCs (chlorofluorocarbons), 133–34, 151
Chakrabarty, Dipesh, 16–18
chaos, 107, 123, 124, 125, 145, 155
*Chaos* (Gleick), 124
*Characterising the Anthropocene* (Macilenti), 88
chemical industry, 136, 151
China, 85, 99–100

# INDEX

Christianity, 91–92, 103, 105
Chthulocene, 15
*The Circle of Reason* (Ghosh), 84
Clack, Christopher, 148–49
Clapeyron, Benoît Paul Émile, 31, 33–35, 36, 42
Clausius, Rudolf, 31, 35–37, 42
climate change, 14; art and, 46–65; global warming effects of, ix–x, xii, 4, 5, 12–13; 27, 45, 130, 149, 150; history of, 6–9; humanities and, 9–10; human understanding of, 18; literature and, 66–89; marginalized peoples affected by, 19, 105, 127, 140, 143; philosophy and, 107–26; politics and, 10–13; population growth linked to, 4; public opinion and, 12–13; religion and, 90–106; scientific consensus on, 1, 11
"Climate Change: An Evangelical Call to Action," 93
climate justice, 95, 131; environmental justice and, 139–40
"The Climate of History" (Chakrabarty), 16–17
closed systems, 37, 40, 45, 123, 171n9
coal, 24, 28, 33, 34, 58, 66; acid rain linked to 151; extraction of, 29, 30, 47, 57, 101; in literature, 67, 71, 83–84; substitutes for, 147, 148; wood supplanted by, 18
"Coal Train" (McPhee), 83
Cobb, John, 101, 103
colonialism, 22, 24, 91–92
*The Columbian Exchange* (Crosby), 14
complexity, 107, 124, 145, 155, 156
*Complexity* (Waldrop), 124
computers, 119–20, 125

*Confronting Environmental Racism* (Bullard), 138
"Confronting the Climate Change Crisis" (Angus), 19–20
Confucianism, 98–100, 101
Confucius, 99, 100
Copernican (heliocentric) universe, 44
Copernicus, Nicolaus, 91, 119
coral reefs, 11
corn, 3, 14
corporations, 24, 46
cows, 3, 14
Crew Company, 69
Cronon, William, 53
Crosby, Alfred, 14
Crutzen, Paul, x, 1–3, 13, 14, 16, 26, 149

dams, 4, 5
Daoism, 98
Davis, Scott, 83
deep ecology, 154–55
deforestation, 105, 150
Democritus of Abdera, 116, *117*
desalination, 63
Descartes, René, 91, 116, 126
desertification, xiii
developing world, xiii, 87, 97, 132, 140, 143, 152
dialectics, 24, 115
Diamond, Jared, 14
Dickens, Charles, 66, 69–72
Dickinson, Emily, 79–80
Diderot, Denis, 27
diesel engine, 45
Dillard, Annie, 66, 81–82
dinosaurs, 17
*Discordant Harmonies* (Botkin), 124

# INDEX

*Discourse on Inequality* (Rousseau), 27
distributive justice, 140
*Dombey and Son* (Dickens), 69–70
domestication, 3, 14
Dow Chemical Company, 136
drought, xiii, 8, 11
*Dumping in Dixie* (Bullard), 138

Earth Charter (2000), 10, 105–6
Earth Ministry, 95
earthquakes, 70, 107, 121
ecocentricity, 127, 129–30
ecofeminism, 16, 87–88
*Ecology and Religion* (Tucker and Grim), 94
"ecology of mind," 154
egocentricity, 127–28, 139
Einstein, Albert, 121–22
electrical grid, 147, 149
electric cars, 62, 85
Eliasson, Olafur, 63, *64*
Elvin, Mark, 99
"The Emerging Alliance of Religion and Ecology" (Tucker), 94
Emerson, Ralph Waldo, 66, 75
Empedocles of Akragas, 116, *117*
*Encyclopédie* (Diderot and Alembert), 27
endangered species, 66
*The End of Nature* (McKibben), 9
Engels, Friedrich, 115
Enlightenment, 26, 45
entropy, ix, 37–38, 40–42, 45
environmental history, 17
environmental humanities, x–xi, xii, 10, 144
environmental justice (EJ), 85, 130–31, 135–36, 137; climate justice and, 139–40

Environmental Justice Resource Center, 138
environmental movements, 65
Environmental Protection Agency (EPA), 4
"Environmental Turn in the Human Sciences" (Sörlin), x
*Envisioning Change* (exhibition), 64
Episcopalianism, 92–93
epistemology, 114
Erie Canal, 53
ethics, 95, 99, 106, 111, 114, 126–43, 154
European Union, 148
Evangelical Climate Initiative, 93
evolution, ix, 44
"The Excursion" (Wordsworth), 67–68
extinction, xiii, 105, 130, 141; increasing rate of, 4, 5
extreme weather, ix, 158n12

*Facing the Anthropocene* (Angus), 20
factory farming, 15
Faith in Place, 95
farmlands, 4
feedback effects, 11, 44
*Feminist Ecologies* (Varney), 88
feudalism, 44
*Fighting Temeraire Being Tugged to Her Last Berth* (Turner), 49, *50*
First Nature, 9
fishing, 150
Fitzpatrick, Paul, 83
*Flight Behavior* (Kingsolver), 86
Flint, Michigan, 137
flooding, ix, 8, 11, 140
food: cost of, 142; shortages of, 8, 152
forests, 4; for carbon dioxide reduc-

tion, 11, 12; destruction of, 105, 150; in Latin America, 96
Forum on Religion and Ecology, 93
fossil fuels, 18; Anthropocene linked to, x, 14, 24–25, 27, 113–14; capitalist profit from, 24, 128; coal, *see* coal; depletion of, 3; divestment policies and, 104; extraction of, 47, 84; formation of, 117; gas, 24, 47; greenhouse gases from, ix, 2, 4, 11, 24–25, 27, 45, 114, 125, 149, 150; importation of, 99; on Native American reservations, 97; oil, 11, 24, 47; in steam engines, 2, 59; substitutes for, 62, 63, 90, 101, 103, 143, 145, 147; for transportation, 85; widespread use of, 134
Francis, Pope, 96
Francis of Assisi, Saint, 91, 96
Friends, Society of, 92
*From Being to Becoming* (Prigogine), 123
Frost, Robert, 66, 80
fuel efficiency, 12–13

Gaia, 155
Galileo Galilei, 91, 126
garbage, 99, 137
Gardiner, Stephen M., 133–35
gas, 24, 47
Gast, John, 56–57
gathering-hunting, 44
gender, 87–89
geocentric (Ptolemaic) universe, 44
geothermal energy, 147
Germany, 148
Ghosh, Amitav, 84
Glacier National Park, 65
glaciers, xii, 65, 97–98, 125

Gleick, James, 124
*The Global Fight for Climate Justice* (Angus), 20
global governance, 135
globalization, 17–18, 97
global warming. *See* climate change: global warming effects of
Global Warming Solutions Act (California, 2006), 10
goats, 3, 14
Goldstein, Rebecca Newberger, 107, 109
Google, 107, *108*, *110*, 112, 148
Gore, Al, 12
grasslands, 4
"great acceleration," 13, 14
*The Great Derangement* (Ghosh), 84
Great Lakes to Ohio and Mississippi Canal, 53
Greek Orthodoxy, 92
GreenFaith, 93, 95
greenhouse gases (GHG), xii, 62, 89, 90, 114, 115, 145, 150–51; climate change skeptics and, 12; early warnings of, 6; historical rise in, 2–3, 14, 24–25, 27, 47, 84–85, 128; projected levels of, 4, 6; reductions sought in, 8–9, 10, 93, 112, 119, 129; scientific consensus on, 7; from transportation, 85–86; unpredictable effects of, 125; worldwide distribution of, 23. *See also* carbon dioxide ($CO_2$) emissions; hydrofluorocarbons (HFCs); methane ($CH_4$); nitrous oxide ($N_2O$); perfluorocarbons (PFCs); sulfur hexafluorides ($SF_6$)
Green parties, 152
"green" science, xi
Griffin, David Ray, 101, 103

# INDEX

Grim, John, 93–94
groundwater, 151
*Guns, Germs, and Steel* (Diamond), 14
Gynocene, 15–16, 87

habitat loss, 130, 141, 143, 151
Haraway, Donna, 15, *16*
*Hard Times* (Dickens), 70–72
Hartshorne, Charles, 101
Hass, Robert, 66
Hawthorne, Nathaniel, 66, 72–74
Hayhoe, Katherine, 105
Healing Our Planet Earth (HOPE) conference, 93
health insurance, 142
heat death, ix, 40–41
Hegel, Georg Wilhelm Friedrich, 115
Heisenberg, Werner, 121–22
heliocentric (Copernican) universe, 44
Heraclitus of Ephesus, 115–16, 117, 119, 120, 125
Hinduism, 98
Hobbes, Thomas, 116, 128
Holocene (interglacial warm period), 3
homocentricity, 127, 128–29, 139
Homogenocene, 15
Honor the Earth, 97
Horkheimer, Max, 153
horses, 3, 14, 46
humanities, x–xi, xii, 10, 144
humanity, 127, 144–56
Hume, David, 27
hurricanes, ix, 142
hydrofluorocarbons (HFCs), 7

ice age, 13
icecaps, ix, xii, 150
"ice car" (Eliasson), 63, *64*

identity, law of, 116
immigration, 24
*An Inconvenient Truth* (film), 12
India: Buddhism and Hinduism in, 98; pollution in, 99
indigenous peoples, 20, 97–98, 104, 105, 127, 135, 140–43
indigenous religions, 92, 94
industrialization, 18, 24, 45, 56–57, 79
inequality, 19, 20, 23, 140, 142
integrative thinking, 154
Interfaith Faith Center on Corporate Responsibility, 93
Interfaith Power and Light Campaign, 104
interglacial warm period (Holocene), 3
Intergovernmental Panel on Climate Change (IPCC), 8–9, 12, 132
internal combustion engine, 45
International Geosphere-Biosphere Programme (IGBP), 7–8, 20
International Society for the Study of Religion, Nature, and Culture, 95
Inuit, 141
Islam, 92, 103, 105

Jackson, Candida Dereck, 104
Jacobson, Mark, 146–47, 149
Jamestown, Va., 15
Jantsch, Erich, 152
Japan, 99
Joule, James Prescott, 39, 42
Judaism, 92, 103
justice. *See* climate justice; environmental justice; procedural justice

Kane, John, 57–59
Kant, Immanuel, 27

# INDEX

Kelvin, William Thomson, Baron, 38–40, 42
Kendall-Miller, Heather, 141
Kepler, Johannes, 91, 119
Keystone pipeline, 97
Khosrowshahi, Dara, 125
Kingsolver, Barbara, 66, 86
Klein, Naomi, 19
Kyoto Protocol, 8, 10

LaDuke, Winona, 97
"The Land Ethic" (Leopold), 129–30
land use, 5, 137
Lao Tzu, 99–100
latent heat, 27
Latin America, 96, 104
Latinos, 138, 140, 142
Lavoisier, Antoine, 27
lead poisoning, 137
*Leaves of Grass* (Whitman), 77
Leibniz, Gottfried Wilhelm, 119, *120*, 162n16
Leiserowitz, Anthony, 12
Leopold, Aldo, 66, 129–30
Leopold, Estella, 66
"The Literature of the Anthropocene" (journal issue), 86–87
Lo, Erin, 85
Locke, John, 128
logic, 116, 117
Lomborg, Bjørn, 12
London and Birmingham railroad, 48
Lorenz, Edward, 122
Los Angeles City Energy Recovery Project (LANCER), 136
Lovelock, James, 155
Lumière, Auguste, 52
Lumière, Louis, 52

*The Machine in the Garden* (Marx), 74–75
Macilenti, Alesandro, 88
Major, Alice, 88
Malthusianism, 20, 152
mammals, 13, 14
Manchester-Liverpool railway, 47
Manet, Édouard, 51
Mann, Charles, 15
*A Manual of the Steam Engine and Other Prime Movers* (Rankine), 42
Marduk (Mesopotamian god), 114
Markoff, John, 107
Marx, Karl, 115
Marx, Leo, 74–75
materialist philosophy, 116
*Mathematical Principles of Natural Philosophy* (Newton), 91
mathematics, 116, 118–21, 124–25
McDaniel, Jay, 103
McGurty, Eileen, 135–36
McKibben, Bill, 9
McNeill, John, 14
McPhee, John, 66, 83–84
meat, 15
mechanics, 119
mechanism, 102, 119–25, 145, 153, 155
MELA (Mothers of East Los Angeles), 136
Melrose, Andrew, 53
"Memoir on the Motive Power of Heat" (Clapeyron), 33–35
Mendoza, Eric, 35
methane ($CH_4$), 3, 7
Mexico, 148
Middlesex Canal, 53
Middletown Woolen Manufacturing Company, 52
Milesians, 114–15

# INDEX

milk, 15
Mill, John Stuart, 129
mining, 24, 29, 30, 47, 57, 84, 101
modernity, 17
Monet, Claude, 51
mono-cropping, 15
*Monongahela Valley* (Kane), 57–59
Monsanto, 86
Moore, Jason W., 21
morals, 99, 101, 131, 135, 146
Morocco, 105
Morrison, Alex, 64
Muir, John, 75
multiculturalism, 127, 130

Naess, Arne, 154
narrative, 26, 46, 56, 57, 76, 92
National Council of Churches, 93
National Indigenous Alliance, 104
Native American Rights Fund, 141
Native Americans, 97, 138, 140, 141–42
native peoples, 20, 97–98, 104, 105, 127, 135, 140–43
"naturalist" philosophy, 114
Newcomen, Thomas, 28–29
Newton, Isaac, 26, 91, 116, 119, *120*, 152
nitrous oxide ($N_2O$), 7, 149
Nordhaus, Ted, 12
North River Sewage Treatment Plant, 137
nuclear age, 13, 14
nuclear energy, 149, 150

oats, 3
Obama, Barack, 86
Obama, Michelle, 86
oceans: acidification of, 11, 105; Japanese pollution of, 99, 151; rising level of, xii, 23; warming of, xiii

oil, 11, 24, 47
"On a Modified Form of the Second Fundamental Theorem in the Mechanical Theory of Heat" (Clausius), 37
"On Changes of Temperature Produced by the Rarefaction and Condensation of Air" (Joule), 39
"On the Dynamical Theory of Heat" (Kelvin), 38–39
"On the Influence of Carbonic Acid in the Air upon the Temperature of the Ground" (Arrhenius), 6
"On the Mechanical Theory of Heat" (Clausius), 37
"On the Motive Force of Heat, and on the Laws Which Can Be Deduced from It for the Theory of Heat" (Clausius), 36
"On the Projected Kendal and Windermere Railway" (Wordsworth), 68–69
"On the Universal Tendency in Nature to the Dissipation of Mechanical Energy" (Kelvin), 40
ontology, 114, 118
open systems, 44, 45, 123, 171n9
*Order Out of Chaos* (Prigogine and Stengers), 123
*Our Common Future* (Bruntland Report), 145
ozone layer, 133–34, 151

paganism, 91
Papin, Denis, 28
Paris Climate Accord (2016), 96
parks movement, 65
Parmenides of Elea, 116, 117, 125
partnership ethic, 127, 131, 146

Pascal, Blaise, 119
"A Passing Glimpse" (Frost), 80–81
Patel, Prachi, 85
Patriarchalocene, 15–16, 60, 87
PCBs (polychlorinated biphenyls), 135–36
"A Perfect Moral Storm" (Gardner), 133–34
perfluorocarbons (PFCs), 7
perspectivist animism, 21
Phallocene, 87
*Philosophical Magazine*, 39
photoelectric effect, 121–22
photons, 121–22
photosynthesis, 150
pigs, 3, 14
*Pilgrim at Tinker Creek* (Dillard), 81–82
plagues, 107, 121
Planck, Max, 121
planets, 117–18
Plantationocene, 15
plastics, 151
Plato, 107, 109, 111–12, *113*, 118, 120, 125
*Plato at the Googleplex* (Goldstein), 107, 109, 111–12
Plotinus, 124, 125
*Poetry and the Anthropocene* (Solnick), 86
Poggendorff's *Annals of Physics*, 36
polar ice, ix, xii, 150
pollution, 150; of air, x, 47, 99, 101, 129; of soil, 101, 151; of water, x, 99, 101, 137
Pontifical Commission on Justice and Peace, 95
population growth, 3–4, 20, 152, 153
poverty, 19, 24, 132, 145
predictability, 107, 119–25, 152

Prigogine, Ilya, 42, 44, 123, 152
*Principia Mathematica* (Newton), 26–27, 119
procedural justice, 143
process, 24, 156; process philosophy, 101–2; process theology, 101, 102
"Process Philosophy and Global Climate Change" (McDaniel), 103
Ptolemaic (geocentric) universe, 44
Pythagoras of Samos, 117–18

quantification, 117–18

"Radical Guide to the Anthropocene" (Klein), 19
radioactivity, 150
railroads, 47–53, 55, 59–61, 68–84
*The Railroad Station in Sceaux* (Manet), 51
"The Railway Train" (Dickinson), 79–80
*Rain, Steam, and Speed* (Turner), 49, 51
rain forests, 150
Rankine, William, 38, 41–42
Raphael, *113*
recycling, 145, 153
*Reflections on the Motive Power of Fire* (Carnot), 32–33
refrigerants, 151
relativity, 122
religion: climate change remediation and, 104–6; Eastern, 98–103; ecology and, 92–98
*Religion, Nature, and Culture* (journal), 95
renewable energy, 11, 62, 90, 101, 106, 119, 143, 147, 150, 153; in developing world, 97; Eastern religions and,

## INDEX

renewable energy (continued) 98; numerical goals for, 13, 148; organized support for, 93
*Republic* (Plato), 111
resource depletion, 101, 150
restoration, 145, 155
rice, 3, 14
*Riprap and Cold Mountain Poems* (Snyder), 81
rock formation, 11
Roman Catholicism, 95–96
*Roughing It* (Twain), 77
Rousseau, Jean-Jacques, 27
Roybal, Maria, 136
Ruddiman, William, 19
Ruether, Rosemary Radford, 96–97
rye, 14

*The Sand County Almanac* (Leopold), 130
Santa Fe Institute, 124
Savery, Thomas, 28
Savi, Melina Pereira, 87, 88
Schori, Katharine Jefferts, 92–93
Schwarzenegger, Arnold, 10–11
Scientific Revolution, 24, 26, 47, 92, 119, 152
sea level rise, xii, 23
Second Nature, 9
self-driving cars, 62, 85
shamanism, 21
sheep, 3, 14
Shellenberger, Michael, 12
Singer, Peter, 132–33
slave labor, 15, 24, 59
Slavocene, 60
Smith, Adam, 27
snowmelt, xii, 23
Snyder, Gary, 66, 91

*The Social Contract* (Rousseau), 27
social ecology, 154
Socrates, 109
soil: depletion of, 87, pollution of, 101, 151
solar energy, 62–63, 97, 101, 105, 146–48
Solnick, Sam, 86
Solutions Project, 147
sorghum, 3
Sörlin, Sverker, x
Spangenberg, Joachim, 99
species migration, xiii, 130
spirituality. *See* religion
squash, 3, 14
stable systems, 123
Standing Rock Sioux Reservation, 97
steamboats, 2, 30, 45, 46, 47, 53, 54, 67–68
"Steamboats, Viaducts and Railways" (Wordsworth), 68
steam engine, 22, 123; Anthropocene initiated by, x, 2, 26, 152; capitalism and urbanization linked to, 24, 46, 52; early versions of, 28, 32; ecological revolution linked to, 46–47; efficiency of, 32, 44; heat source of, 33, 34; iconography of, 49–61; second law of thermodynamics and, 37–38, 45; in textile industry, 48–49; Watt's, x, 2, 26, 27, *30*, 31, 42, 45
Steffen, Paul, 14
Steinbeck, John, 66
Stengers, Isabelle, 123
Stephenson, George, 47
Stevens, Lara, 88
Stockton and Darlington Railway, 47

# INDEX

Stoermer, Eugene, x, 1–3, 13, 16, 26
"The Stone Garden" (Snyder), 81
StopGlobalWarming.org, 93
Stott Park Mill, 49
Styrofoam, 151
sulfur hexafluorides ($SF_6$), 7
sustainability, xii, 62, 99, 101, 106, 126, 131, 143, 144–56
SUVs (sport utility vehicles), 13

Taoism, 98–100
*Tao Te Ching* (*The Way*; Lao Tzu), 99–100
Tate, Peta, 88
technology, 9, 23, 24, 74, 76, 80, 114, 119, 125
telegraphy, 56–57
textile industry, 48–49, 53
Thales of Miletus, 114
thermodynamics, 24; classical, 42, 123; far-from-equilibrium, 42, 44, 45, 123; as field of science, 38–42; first law of, 35; second law of, ix, 31, 35, 36–38, 40, 44–45
*This Changes Everything* (Klein), 19
Thoreau, Henry David, 66, 75–76
350.org, 6, 103
"To a Locomotive in Winter" (Whitman), 77, 79
tornadoes, ix
toxic waste, 138, 142, 151
"Toxic Wastes and Race in the United States" (report), 138
*Transforming Environmentalism* (McGurty), 135–36
Trexler, Adam, 86
Trump, Donald J., 96
tsunamis, 107, 121
Tucker, Mary Evelyn, 90, 93–94

Turner, Joseph, 49, *50*
Twain, Mark, 77
"Two Rivulets" (Whitman), 77
Tycho Brahe, 119

uncertainty principle, 122
*Uncovering the World That Columbus Created* (Mann), 15
Union Pacific Railroad, 55
Unitarianism, 92
United Church of Christ, 138
United Nations, 10–11; Environmental Programme of, 8
University of Glasgow, 29, 38
unpredictability, 107, 119, 121–23, 125, 145
urbanization, 24, 46, 52, 65
utilitarianism, 129, 139

Varney, Denise, 88
vegetables, 15
Viveiros de Castro, Eduardo, 20–21
volcanoes, 107, 121
Voltaire, 27

*Walden* (Thoreau), 75–76
Waldman, Anne, 86
Waldrop, Mitchell, 124
water: cultural and spiritual uses of, 142; desalination of, 63; pollution of, x, 99, 101, 137; supply of, xiii, 63, 141, 151
watermills, 28
Watt, James, x, 2, 26, 27, 29–30, 42, 45
*The Wealth of Nations* (Smith), 27
*Welcome to the Anthropocene* (Major), 88
*West-Running Brook* (Frost), 80

# INDEX

*Westward the Star of Empire Takes Its Way* (Melrose), 53, *55*
wetlands, 4
wheat, 3, 14
White, Lynn, Jr., 91
Whitehead, Alfred North, 101
Whitman, Walt, 66, 77–78
wilderness, 53, 65, 66, 76
wind energy, 28, 97, 101, 146–48
women, 127, 131, 146; in developing world, xiii, 87; male domination of, 16, 155; as railroad workers, 60, *61;* sustainable livelihood and, 145
Wordsworth, William, 66, 67–69
"The Young American" (Emerson), 75
*Your Mobile Expectations* ("ice car"; Eliasson), 63, *64*

Zen Buddhism, 98

Printed and bound by CPI Group (UK) Ltd, Croydon, CR0 4YY

15/12/2024

14612142-0001